Rapid Reliability
Assessment of VLSICs

Rapid Reliability Assessment of VLSICs

A. P. Dorey, B. K. Jones, A. M. D. Richardson, and Y. Z. Xu

University of Lancaster
Lancaster, United Kingdom

PLENUM PRESS • NEW YORK AND LONDON

Library of Congress Cataloging-in-Publication Data

Rapid reliability assessment of VLSICs / A.P. Dorey ... [et al.].
    p.   cm.
  Includes bibliographical references.
  ISBN-13:978-1-4612-7879-5   e-ISBN-13:978-1-4613-0587-3
  DOI:10.1007/978-1-4613-0587-3

  1. Integrated circuits--Very large scale integration--Testing.
2. Integrated circuits--Very large scale integration--Reliability.
I. Dorey, A. P.
TK7874.R37 1990
621.39'5--dc20   89-72202
                                                                CIP

© 1990 Plenum Press, New York
Softcover reprint of the hardcover 1st edition 1990

A Division of Plenum Publishing Corporation
233 Spring Street, New York, N.Y. 10013

FOREWORD

The increasing application of integrated circuits in situations where high reliability is needed places a requirement on the manufacturer to use methods of testing to eliminate devices that may fail on service. One possible approach that is described in this book is to make precise electrical measurements that may reveal those devices more likely to fail. The measurements assessed are of analog circuit parameters which, based on a knowledge of failure mechanisms, may indicate a future failure.

To incorporate these tests into the functional listing of very large scale integrated circuits consideration has to be given to the sensitivity of the tests where small numbers of devices may be defective in a complex circuit. In addition the tests ideally should require minimal extra test time.

A range of tests has been evaluated and compared with simulation used to assess the sensitivity of the measurements. Other work in the field is fully referenced at the end of each chapter.

The team at Lancaster responsible for this book wish to thank the Alvey directorate and SERC for the necessary support and encouragement to publish our results. We would also like to thank John Henderson, recently retired from the British Telecom Research Laboratories, for his cheerful and enthusiastic encouragement. Trevor Ingham, now in New Zealand, is thanked for his early work on the project.

We also enjoyed the support of many manufacturers; without their support the project would have been impossible. For obvious reasons certain details of devices cannot be published, the aim of the book being to describe the measurement techniques in the hope that others will apply the ideas to their products where high reliability is needed.

FOREWORD

CONTENTS

Chapter 1

INTRODUCTION TO VLSI TESTING

1.1. THE PROBLEM

Digital, very large scale integrated circuits (VLSICs) are used widely. In many applications the incorrect function of the circuit upon installation, or the malfunction or failure during use, are inconveniences which are often detected during the early operation or burn-in period of the system in which the circuit is used. However, in some applications where the replacement cost is high or the consequences of failure are serious, highly reliable devices of high quality are needed. Examples of such uses are in satellites, undersea cables, remote stations, manned space vehicles and for military systems.

This book describes methods of assessing the quality and reliability of VLSICs using sensitive, non-destructive, electrical measurements. A selection of tests has been devised and evaluated on relatively small scale integrated digital circuits made with CMOS technology. The tests have been shown to be valuable both for reliability assessment and for determining the location of any failed component within the circuit. Suggestions are made for the techniques which may be used to implement these tests in an industrial environment.

1.2. RELIABILITY TESTING

The present methods of testing integrated circuits (ICs) are incomplete. The tests are for operation, for conformance to manufacturing specification, for quality of manufacture and for statistical lifetime.

The basic test for operation is the 'functional test'. In this test the digital IC is tested at its rated voltage and speed to ensure that it performs to specification. For the logic aspect of the circuit this means that the correct output signals are observed for all possible combinations of input signal (input vector) and for sequential logic, all possible combinations of internal node status or transitions of internal node status. In practice such a full test would be impractical and very lengthy for even a medium scale integrated circuit (MSIC) so assumptions are made about the likely faults and failure modes, and a reduced sequence of input vectors is used. For most commercial applications a much reduced sequence is used. Even the full functional test gives only information that the device is working to its specification at the time of testing. This is the minimum standard required in order that the device may be used in a critical application. The test gives no information about the quality of manufacture other than pass/fail. It also gives no information about the device's possible reliability or lifetime in use.

The next extension to the functional test is to provide a burn-in period for the completed device before a final functional test. During the burn-in time the device is exercised in as full a manner as possible so that currents flow in all signal paths and the input and output signals and internal nodes are taken through all possible combinations. This operation is performed at a moderately high temperature on the assumption that any gross weakness in the device will rapidly be extended significantly to produce a functional failure. Burn-in is a well established procedure which simulates a long period of normal use and detects infant mortality. For greater confidence, when the circuit is incorporated into a system this is then itself subjected to a similar burn-in period. This test is non-destructive, at least for the 'good' devices, and subjects the circuits to little excess stress. Although this burn-in procedure is useful, it gives no information about the possible device lifetime or whether any particular batch has been well fabricated and will have a longer statistical lifetime than another batch. The procedure only picks out those devices which have been particularly badly made or which possess a gross flaw.

For high reliability a circuit should have a conservative design using a well established fabrication process. For example, the design would have well spaced components, low current densities and low device temperatures due to self-heating. The fabrication process would have well verified processing steps and operations which have been shown, for this or for other designs, to exhibit high reliability. The high reliability, or long statistical lifetime, is best established by the analysis of a large number of devices run in a normal operation. It is apparent that both a conservative design and a well established fabrication process are not compatible with the rapid advances being made in electronic circuits with new and faster circuits of smaller element size and larger scales of integration. By the time sufficient device-years of use have been developed the device is likely to be obsolete.

While the general principles above are desirable, the emphasis on reliability assessment is to ensure high care in the manufacturing process, to carry out accelerated life tests on samples and to investigate the detailed cause of actual failure.

The care in manufacture is controlled by strict conformance to the high specification of the fabrication process. Frequent checks are made on the quality of the materials and process operations. Occasional non-destructive checks can be made on the individual die using microscopy or electrical probing and more detailed checks of the processing can be made on specific test chips or test devices. In this way good reproduction of an established process can be maintained. In many ways this is a straightforward extension to standard quality control and assessment to ensure high yield. It improves quality with the implication of increased reliability. Although the tests and checks performed are designed to be innocuous and non-destructive, the increased handling and number of operations are likely to cause unwanted problems.

Accelerated life testing is a standard technique for assessment of reliability or statistical lifetime. To obtain a value for the statistical lifetime of a particular type of device, a very large number of devices need to be operated under normal conditions for a long time in order to obtain a statistically significant number of failures. Modern devices can be very reliable. Life testing under normal operating conditions is impractical because of the time taken, the numbers of devices needed and the cost of the exercising circuits involved. It is normal practice to decrease the time and sample number considerably by increasing the stress magnitude. Thus in a radiation stress the particle flux is increased greatly above the level expected in use to enhance the failure process. This is justifiable provided that the failure mechanism is not altered: for example, a different failure mechanism may only occur at high fluxes or there may be a radiation rate dependence of the mechanism. This basic assumption is a major problem since experiments over a wide range of acceleration factors are not possible for practical reasons and the only recourse is to a comparison of the failure modes in normal usage and in the accelerated test. In practice the number of devices tested in accelerated tests is very small since the equipment is expensive; hence the statistical accuracy of the experiment is not very good.

The most common accelerated life test is that done at elevated temperatures. A small sample is operated at its normal operating electrical conditions and excercised through a wide sequence of test vectors at an elevated temperature. It therefore experiences conditions similar to normal operation except for the temperature. Most failure mechanisms are thermally activated so that the failure rate varies as $e^{-(E_A/kT)}$, where E_A is the activation energy of the process, k is Boltzmann's constant and T is the absolute temperature. A change in the temperature thus changes the failure rate. In order that the same failure mechanism operates at the elevated temperature, other competing processes should not have similar values of E_A.

Accelerated life testing can be a valuable method for evaluating the reliability of a process and to produce failures rapidly which may indicate the least reliable parts of the process. However, the procedure takes some time, is inherently destructive, is statistical in nature, investigates a small number of samples and does not assess the reliability or quality of the actual devices to be used in the system which requires high reliability.

What is needed is a simple, fast, reliable, non-destructive test which can be performed on the finished device to evaluate its quality of manufacture and its likely reliability. The test should be generally applicable since processes and technologies change over the years. The test should also be very sensitive so that one weak or defective element of the IC can be detected even on a VLSIC. The test cannot measure reliability as such but can only measure the quality of manufacture of the device. There is likely to be a strong connection between these attributes, and high quality of processing is already assumed to be a requirement for high reliability devices. The absolute numbers derived from the test, and to be used as a measure of the quality or reliability, will vary between designs and more between processes so that calibration will be needed on an absolute scale. However, for one device type a relative scale of quality can be established easily.

It is apparent that since the required test is one of quality there may also be a use for it in production quality control and yield improvement, as well as for vendors' incoming quality control.

1.3.   THE CMOS PROCESS

The complementary metal–oxide–silicon (CMOS) process was chosen as the technology for this study because it is very widely used now and is predicted to be used extensively for some time in the future. The basic technology is simple with both n-channel and p-channel MOS transistors. Since the type of doping of the substrate allows only one of these transistors to be fabricated directly, there is selective doping of 'wells' in the substrate in which the other type is fabricated. The MOS technology in its simplest form requires few processing steps and hence masks. More complex processing is needed for CMOS and for more refined and smaller geometry versions of the general family.

The basic circuit element is a series-connected complementary pair of transistors between the power supply voltages, V_{dd} and V_{ss}. The input signal is applied to both of the gates together. As one transistor turns off the other turns on, so that the binary logic states are with the output, at the midpoint of the series combination, connected to either the upper or lower power supply potential. A critical requirement therefore is that the threshold voltages of the two transistors be matched to each other and to the supply voltage so that there is a smooth transition as the input voltage signal changes level. When the output is in either of the two binary states no current flows through the complementary transistor pair. The power supply only has to supply current during the transition. Some current flows when both transistors are turned partially on, but this is very small for a sharp transition, and the largest contribution comes from the charging and discharging of the capacitance connected between the output of that gate, the common point between the transistors and earth. The size and length of this charging pulse depend on the size of the capacitance and the current sourcing capacity of the transistor supplying the current.

1.4.   FAILURE MODES AND MECHANISMS IN CMOS

The CMOS technology suffers from many of the reliability problems of other technologies. In digital circuits the basic fault conditions are with an output or internal node stuck at logic–0 or logic–1. The reason for this state may be simply

4

expressed as the current in a part of the circuit being too high or too low. There may be a simple short circuit or open circuit such as caused by a low impedance bridge between two tracks or an electromigration track continuity failure, but the cause could also be a leaky junction, a low gain transistor or a deviation in a threshold voltage.

The general problems of MOS VLSI reliability and yield have been discussed by Woods (1986). In use, the technology suffers the general problems of all ICs due to corrosion when the encapsulation is not perfect, electrostatic discharge when used in a poor system environment and soft errors due to isolated ionizing radiation incidents. Because of the small charges involved in CMOS the latter problems can be serious if precautions are not taken.

The normal fabrication inadequacies also cause problems. These include scratches, chemical impurities, dust and blemishes, photolithography misalignment and misregistration and incorrect processing times and temperatures. The effect of these are many and various, but they will all produce a finished circuit with less than ideal properties if one assumes that the fabrication process had been optimized. Although the results of these processing problems will, by definition, be bad overall, it is not necessary that the deficiency will be immediately apparent. Although an incorrect threshold voltage will cause poor operation, it is possible to imagine some process errors which could improve some properties at the expense of others. For example a poor registration may result in narrow interconnect line widths. This would decrease the interconnect stray capacitance and speed up the charging time during a transition and hence increase the speed of operation. However, the current density would increase and the narrow lines would be prone to premature failure due to electromigration. This analysis is very simple and in most cases there is likely to be immediate deterioration in other properties also.

The main long-term failure mechanisms occur in the oxide, the metal interconnects and the p–n junctions. Although these wear-out failure mechanisms may be accelerated or enhanced by poor fabrication they are basically controlled by the quality of the design.

The gate oxide plays an important role in MOS technology so that it has to be made to a high quality and not deteriorate. A flow of oxide current will eventually produce gradual wear-out. A high electric field produces a more sudden breakdown with large and destructive current flow. Since MOS devices operate by charge control, the presence of unwanted charges in the gate oxide can have a serious effect. Thus mobile ionic charges within the oxide, fixed oxide charges and charges in electronic states at the oxide–semiconductor interface can cause a great decrease in performance. These charges can be produced by contamination, poor processing and damage by high energy carriers. A particular form of degradation in MOS devices is that due to hot carriers which gain considerable kinetic energy in the high channel fields and can give up this energy in damaging the lattice or creating electron–hole pairs which result in excess currents.

The metallization problems are mainly associated with electromigration which is a physical displacement of the metal atoms caused by the momentum of the carriers in the current. This effect obviously increases with the electrical current density. It is highly dependent on the microscopic structure of the metal and the geometry since the process is that there is a net atom flux away from some areas and therefore the metal cross-section will reduce. Since the current density consequently increases, the process accelerates and failure is rapid once it has begun. The same basic phenomenon also occurs at metal–semiconductor contacts.

Deterioration in the quality of the p–n or metal–semiconductor diode junctions is normally apparent as an increase in the leakage current in reverse bias. This can be due to a gradual increase in the number of defects and resulting electron traps in either the bulk depletion region, in the oxide or at the oxide–semiconductor interface where the depletion region intersects the surface. In more severe cases irreversible breakdown of the junction results in a short circuit.

The importance of these basic mechanisms of failure in future devices can be seen by considering the effect of scaling all the linear dimensions of the process by a factor $K > 1$ to reduce the size.

Table 1.1. The Impact of Idealized Scaling on Device Dimensions
and Material Parameters

Scaling factor $K > 1$		Ideal scaling
Dimension		
Channel length	L	$1/K$
Channel width	W	$1/K$
Gate oxide thickness	t_{ox}	$1/K$
Junction depth	X	$1/K$
Contact length	L_c	$1/K$
Line width	d	$1/K$
Metal thickness	t	$1/K$
Material parameter		
Substrate doping	N_A	K

Table 1.2. The Impact of Idealized Scaling on Device Electrical
Parameters

Scaling factor $K > 1$		Constant voltage	Scaled voltage
Supply voltage		1	$1/K$
Device current		$K \to 1$	$1/K^{0.5} - 1/K$
Average current per contact or line		K	$1/K^{0.5}$
Current density	J_M	K^3	$K^{1.5}$
Oxide capacitance	C_{ox}	$1/K$	$1/K$
Oxide storage charge	Q_{crit}	$1/K$	$1/K^2$
Oxide field	E_{ox}	K	1
Junction capacitance	C_1	$1/K^{1.5}$	$1/K$
Junction storage charge		$1/K^{1.5}$	$1/K^2$
Power dissipation per gate		K	$1/K^{1.5}$
Gate delay		$1/K^2$	$1/K^{1.5}$
Delay $\times$ power		$1/K$	$1/K^3$

Tables 1.1 and 1.2 are based on those by Woods (1986). In Table 1.1 the scaling is described with all dimensions scaled and the doping density scaled to compensate. Table 1.2 shows the effect of the scaling on the electrical and stress quantities. In the simplest form of geometric scaling the voltage is kept constant since the circuit is then still compatible with earlier technology. This is shown in the constant voltage column, and the result is greatly enhanced current densities and electric fields. To reduce these stresses it is possible to set up a new system standard at a lower supply voltage which is assumed to scale by the same factor. This is shown in the scaled voltage column. These two scaling methods have different relative advantages and disadvantages but both increase reliability problems.

The significant features of this estimation are that current densities in the metal interconnects and contacts increase and so enhanced electromigration is expected. The effect of unwanted oxide charge is enhanced as indicated by the critical value Q_{crit}. Also the self-heating of the chip increases. This produces a secondary enhancement to degradation processes which are mostly thermally activated.

1.5. OUTLINE OF THE PROJECT

1.5.1. Requirements for the Tests

The aim of the investigation was to develop and evaluate electrical tests that could be made on completed, encapsulated devices to determine whether they were well made and were hence likely to have high reliability. Although the tests could probably also be made on the chip before encapsulation, it is assumed that simple electrical tests have been made at that stage to determine which devices are acceptable and worth encapsulating. The tests should be totally non-destructive and hence at a very low stress level so that no extra stress would be involved. With such tests it is hoped that the number of investigations of the chip during the various stages of its processing could be reduced and hence there would be less likelihood of damage to the chip during the testing process.

These tests for incipient failure are not likely to detect circuits with poor design rules or poor processes, for example, designs with inadequate metallization track width and spacing and metallurgical combinations which are prone to interdiffusion or corrosion. The tests are therefore no substitute for some form of reliability assessment of the process design and layout rules by such means as accelerated life testing, enhanced stressing and the investigation of the mechanism by which real devices have failed in normal use as well as in accelerated life tests. The tests are designed to detect early failure during normal operation rather than wear-out failure or infant mortality which could be removed by a burn-in procedure.

The tests may be considered as ways of evaluating a processed batch of devices to ensure that the manufacturing quality has been up to some achievable norm or, for higher quality selection, whether the devices are better than 90% (say) of what is normally achieved. The tests would therefore have value in quality assurance (QA) on a routine basis, although as they are more complex than the normal simple QA tests perhaps they would only be used on a sample basis. These tests for quality of manufacture, rather than the quality or potential of the process, would have great value in the extension of standard processes to new designs and especially to application-specific integrated circuits (ASICs) where very small runs of particular designs are produced, no detailed test procedure can be devised, and there will be little history, statistical information of quality and failure or the possibility of detailed reliability assessment for that particular design.

The tests should be able to detect a departure from the ideal process specification, so that the whole chip has substandard performance, and also whether one component of the whole chip is weak or defective. The former type of departure from the optimum is likely to be detected by standard process control procedures such as the evaluation of the properties of a test chip. The latter is likely to be caused by a point blemish, such as a dust particle, which may not be large enough to cause the incorrect operation which would be detected by a full functional test. It is this type of circuit weakness, which may be the cause of failure after some period of use, that is so difficult to detect normally and is the main focus of this investigation.

1.5.2. Possible Measurements

The tests that may be used for this objective must be non-destructive and at a low level so that the device is not unnecessarily stressed beyond the normal specified ratings.

The problem with such tests on digital VLSIs is that there are very many components on the chip (of which only one may be defective or weak) and there are only a very few external contacts to the circuit. Direct contact to a defective component is not possible and some form of discrimination to identify the effect of the weak component is necessary. This implies, in general, that the test should be very sensitive to defective properties.

The external contacts to the circuit may be classified as input contacts, output contacts and the power supply. Taken by itself an input connection or an output connection can only give information about the properties of that section of the circuit, namely the complementary transistor, series-connected pair at the input or output. Measurements on these are fairly unproductive. The input impedance describes the gate properties of the input stages which only sample a very limited property of a few components. Similarly, the output impedance gives the gain factor of the output transistors, which is also very limited.

Of more interest is the combination of the input or output terminal and the power supply. Again the output stage will be in either the logic–0 or logic–1 state so that little more information may be obtained via the output terminal than from an output impedance or current sourcing ability of the output transistors. However, the input stage can reveal more information. If the input potential of one input connection is gradually varied between the power supply potentials V_{ss} and V_{dd}, then the input stage is slowly switched between its two logic states and the supply current changes as one device is gradually turned on with the other fully on. All other gates only contribute a current during the short time when they make a transition. This measurement is described in detail later in Section 3.3.6. It gives values of the gain factors (K_n, K_p) and threshold voltages (V_{Thn}, V_{Thp}) separately for the two input transistors. A sample taken in this way of all inputs gives a good measure for the average chip threshold and gain parameters of all the devices over the chip and hence the variation over the chip.

The I–V characteristic of the whole circuit with the gates and internal nodes set at some fixed logic pattern is also very revealing. Since no current flows in an ideal CMOS circuit when it is not making a transition, this current represents the leakage current in all the p–n junctions, gate oxide capacitors and transistor channels that are being subjected to a significant reverse voltage. Other junctions may be in series with these junctions and in forward voltage bias but, therefore, will not make a serious contribution. By changing the input signal combination (input vector) and the sequence of these vectors, the voltage distribution among the internal nodes can be changed and other combinations of junctions subjected to reverse bias. The size of the leakage current is known to be a sensitive function of the defect properties of a junction and is controlled by impurities and isolated electronic states in the bulk or surface depletion region. The shape of the leakage current I–V curve is indicative of the problem such that some diagnosis is possible. By means of a suitable selection of input vectors, high or low discrimination is possible. For example, it may be possible to perform a very crude but fast experiment in which in one measurement half the junctions are reverse biased and in the second measurement the other half are. A slower test, but one with higher discrimination, may be to put all the junctions of the circuit into reverse bias one at a time so that the change in the leakage current between the static condition and its neighbors is a measure of the leakage of that junction.

As well as the magnitude and voltage dependence of the leakage current being sensitive to the defect concentration, there is good evidence that this leakage current becomes noisy when it is caused by defects. At constant voltage bias the current fluctuates about its mean value. In a good quality device this will only show shot noise, which has a constant spectral density with frequency. However junctions with defects of almost any sort show an enhanced noise intensity at low frequencies, and this often has a $1/f$ spectrum. The magnitude of this excess noise is a measure of the defect property of the junction. This noise can be studied in the leakage current at each input vector. However, since the excess noise is usually only significant below a frequency of about 1 Hz this measurement is inherently slow, although in principle a sampling technique could be used to overcome the problem.

The dynamic properties of a digital IC have great potential for revealing the detail of the internal operation of the circuit. As well as the problem of discrimination of one defective component within the whole circuit there is another problem, signal self-healing, which is inherent in binary digital circuits. Consider a signal passed along a sequence of binary gates of which one is weak. This weak gate may only operate marginally, it may have a poor threshold voltage or gain factor or its output levels may not be correct. However, once its output signal passes the threshold voltage of the following gate the signal has been restored to its full binary logic levels. The defective signal from a weak gate does not propagate but is corrected by the following gate. To detect the defective gate therefore requires a very detailed test.

The defect, or weakness, in a gate can be revealed by reducing the supply voltage or increasing the signal rate until the gate ceases to be able to operate the following gate. The effects may be due to substandard threshold voltage or poor current sourcing ability and may be related, but a test at a combination of low voltage and high speed is likely to reveal the problem.

The problem of discrimination may be solved in two ways. If a particular serial chain of gates is set up between an input and an output contact then the signal will not propagate if the signal speed is increased or the supply voltage is decreased beyond the operating value of the weakest gate. Different signal paths can be set up through the circuit to detect and perhaps locate any particularly weak element. In practice, an easy method of performing this test is to set the supply voltage to a low value, near the operating limit, and increase the signal frequency until the signal does not propagate. This is the frequency of malfunction.

The discrimination to detect a single weak device and to detect its properties may also be made by monitoring the current flowing from the power supply as the device is exercised. Only the leakage current flows when the device is not switching. If a complementary transistor pair in a gate makes a transition, a current pulse flows to charge the capacitor at the output node. In principle, therefore, discrimination can be obtained by switching, ideally, one gate at a time and observing the resulting current pulse. In practice, several nodes may switch at some input vector transitions but high discrimination exists and often the individual gate transitions are not exactly simultaneous so that the current pulse will consist of several distinguishable constituent pulses. The size and shape of these current pulses give considerable information about the constituent transistors. In particular, the length of the pulse is a measure of the speed of switching and hence gives similar information as the frequency to malfunction test described earlier.

A further property of the current pulse can indicate that the components in the particular current path which is being exercised are defective in that the current will show excess $1/f$ noise. As before this noise is indicative of the presence of defects. The fluctuation in the current results in amplitude modulation of the current pulse which may be demodulated for detailed study.

Again the excess noise measurement is intrinsically slow, although sampling techniques may speed the process with increased complexity in the apparatus. The main problem with all the current sampling experiments is that in order to obtain detailed information the shape of the current pulse is needed, hence a recording system is required which is much faster than the speed of the digital circuit under test. To investigate the value of such tests we have used a fast sampling system and slow CMOS circuits, but this approach is probably not suitable for industrial use especially when the next generation of ICs will be faster and thus render the equipment obsolete. However, if the tests are found to be of value then simpler methods of measuring the parameter of interest are often possible.

1.5.3. Generation of Defective Samples

The project involves the investigation of sensitive tests to detect a single weak component within an IC. The tests have to be developed and their performance verified. Modern ICs are usually well made. Some are given a functional test. Therefore supplies of known defective devices are not readily available to verify the tests.

An ideal experiment would involve the testing of a large number of devices before they are put into regular service and then comparing the failure statistics with the test predictions. This experiment would require a large number of devices and a long time. The time could be reduced by performing accelerated stress tests but this would still involve a large number of devices in expensive facilities since few devices will contain initially weak components.

The methodology adopted was to study small batches of devices and to stress them over a period of time so that defects appeared and the devices eventually failed. Initially, and at intervals during the stress period, the tests were performed to determine whether they could detect the defective devices earlier in the stress process than the normal indicator, which is failure on a functional test. The essence of the methodology is the assumption that the type of defect accentuated or introduced by the progressive stress period is typical of the type of defect occurring in a newly manufactured device going into normal usage. While this assumption may not be strictly correct, it is likely that very similar types of defect will be produced.

1.5.4. Stresses Used

The main stress chosen to produce the progressive deterioration was temperature with the devices exercised at a low frequency during the stress. One batch was encapsulated in plastic and stressed at a high temperature. This was likely to produce extra failures above these found for the other devices in ceramic encapsulations.

In a separate experiment some devices were stressed with X-radiation with the same testing procedure at intervals during the stress.

1.5.5. Device Simulations

To obtain a full understanding of the operation of the devices the test results of the circuits were simulated using the SPICE circuit simulator package. The simulation reproduced the frequency to malfunction test and also the current transient test results.

The main value of the simulation was to perform a sensitivity analysis on the various failure modes. Different circuit weaknesses were inserted into the simulation and the simulated test result obtained. In this way the size and type of weakness could be estimated for the experimental results obtained.

1.6. CHOICE OF DEVICES

The CMOS process technology was chosen for the study since it is very commonly used and is likely to continue to be used for some time, although with improved and modified processes.

For general applicability and acceptance the devices chosen should be readily available and well known. A suitable industry-standard set of devices are the 4000B series of circuits. These are small and medium scale. Although the design is rather old, the processing is likely to be at a high industrial level such that good yields are possible. This choice has the advantage for the initial experiments, although not for the final application of the ideas, that the devices are large and relatively slow so that experiments on them are fairly easy to carry out. These differences between the initial test devices and modern, high-speed VLSI devices must be remembered when the tests are applied in the industrial context.

In the initial stages of the investigation tests had to be developed for CMOS ICs which were practical, sufficiently sensitive and indicated that they could detect weak or failed devices. At this stage several tests were devised and evaluated before the most promising tests were selected. Most of the tests evaluated were based on those already carried out in some form on discrete devices.

For the initial investigations simple devices were needed. The 4013 dual D-type flip-flop was used since it is relatively small scale and the fact that it was dual allowed a comparison of failure or weakness in two distinct halves of the chip with few common components. This gives very direct evidence as to whether the failure is local to one part of the chip due to a blemish, or due to deficiencies across the whole chip such as when process control is not good.

The next scale of integration was the 4014 8-bit shift register. However, for extension to real devices the much larger scale device was needed but one with sufficient similarity to previous devices that the earlier experience was of value. The circuit simulation investigation had been shown to be useful and it was thought necessary to have control over the device design and to have good knowledge of the simulation SPICE parameters. A semi-custom design was chosen. With the great popularity of such circuits, testing of small batches will be necessary and will be a problem since the time and effort in designing specific tests for functionality and reliability for each design will be large. The semi-custom chip was designed in two parts with a 4×4-bit multiplier followed by 8-bit shift register. This had the advantage that the multiplier uses combinational logic which had not been tested and also the chip contains a shift-register which was familiar from the 4014. Similarities and differences between the two versions of the shift-register could be evaluated. Details of the designs will be found in Section 2.2.

1.7. OUTLINE OF THE BOOK

Chapter 2 describes in more detail the properties of the devices studied and the methods used in the circuit simulation process. The results of the sensitivity analysis using the simulation are discussed.

The actual experimental methods and apparatus used for the test and stress circuits are described in Chapter 3 together with the typical results obtained for normal specimens.

The results obtained during the stress period as devices gradually failed are outlined in Chapter 4 together with a comparative analysis of the value of the different tests. There is a description of typical test signals given by abnormal devices and a brief account of how the tests may be used to locate a defective component and to identify the manner in which it is defective.

Although most description is given of the tests and their value, some thought has been given to the actual implementation of the tests in a production environment. This concept is described in Chapter 5 together with the test strategies which are recommended. The description for implementation is speculative since each particular application would need separate treatment. Finally, the conclusions of the study are given in Chapter 6 together with other possible directions of study.

REFERENCES

General

Healy, J.T., 1980, *IEEE Test Conference*, 502-6.
Xu, Y.Z., 1986, *PhD Thesis*, University of Lancaster.

Failure Modes and Mechanisms in CMOS

Amerasekera, E.A., and Campbell, D.S., 1987, "Failure Mechanisms in Semiconductor Devices", Wiley, Chichester.
Brambilla, P., Canali, C., Fantini, F., Magistrali, F., and Mattana, G., 1986, *Microelectron. Reliab.*, **26**: 365-84.
Brambilla, P., Fantini, F., and Mattana, G., 1983, *Microelectron Reliab.*, **23**: 761-5.
Edwards, D.G., 1980, *IEEE Test Conference.*, 407-16.
Fantini, F., 1984, *Microelectron. Reliab.*, **24**: 275-96.
Fantini, F., and Morandi, C., 1985, *IEE Proc.*, **132 G**: 74-81.
Holton, W.C., and Cavin, R.K., 1986, *IEEE Proc.*, **74**: 1646-68.
Stojadinovic, N.D., 1983, *Microelectron. Reliab.*, **23**: 609-707.
Vincoff, M.N., 1974, *IEEE Trans. Reliability*, **R24**: 255-9.
Woods, M.H., 1986, *IEEE Proc.*, **74**: 1715-29.
Woods, M.H., and Euzent, B.L., 1984, *Tech. Digest Int. Electron Devices Meeting (IDEM)*, 50-5.

Chapter 2

THE DEVICES STUDIED
AND THEIR SIMULATION

2.1. INTRODUCTION

As part of the general investigation of rapid reliability assessment the devices being studied experimentally were also simulated using a standard circuit simulation package, PSPICE (Microsim Corporation), which is a version written for use on the IBM personal computer. The simulations gave the size and shape of the waveforms occuring in the circuits being studied and the effect of different conditions within the circuit on these waveforms.

These simulations were very valuable in the study for several reasons. Firstly, by comparing the simulation results with those obtained experimentally on normal devices we obtain confirmation that we understand the operation of the device and that we can perform the simulation accurately with suitable device parameters. This confirmation between the experimental and simulated output signals can then be extended to simulation of the operation of the inaccessible internal gates, transistors and nodes. This enables a better understanding of the operation of the circuit and interpretation of the results of the experiments. For example, it may be possible to determine which sections of the circuit have the longest time delay or to determine the relative switching times in different parts of the circuit which may reveal some internal race conditions. For instance, we showed that the charging of one internal node capacitance of the semi-custom multiplier chip dominated the speed of response of the circuit.

The second use of the simulations was to study the properties of circuits which showed an abnormal behavior either when purchased or as a result of stressing. Experimentally abnormal devices show changes in one or more of their measured electrical properties. By changing the device parameters, such as threshold voltage, gain factor or leakage current, in the simulation the resulting change in the circuit performance could be observed and identified. A relationship between the changes in the simulated and experimentally observed performance gives a strong indication of the device parameter which is causing the observed effect. This simulation also allows determination of whether the effect is in one transistor or gate, as in a point blemish such as electromigration failure, or in all transistors on the chip, as caused by an overall threshold voltage change during irradiations. A detailed simulation of the magnitude of the observed effect resulting from an increasing magnitude of the supposed defect enables a sensitivity analysis to be performed so that it is possible to determine how carefully the device design and processing has to be carried out in order to produce devices within specification, and how far the parameters of individual transistors, or the chip as a whole, may deteriorate during use or stressing before the circuit properties fall below specification.

Finally, a detailed simulation of the deterioration of one parameter of one transistor will produce results which can be compared with those obtained experimentally. In this way confirmation may be obtained of the type and location of the particular fault observed in a circuit which was observed to fail slowly in the stress experiments.

2.2. DETAILS OF THE DEVICES STUDIED

The particular IC devices were chosen for study to present a graded increase in difficulty and complexity as the investigation progressed. They all employed CMOS technology and were available from normal commercial sources and so represented typical devices. Similarly, in order to demonstrate that the results were general the industry-standard 4000B series devices were used.

CMOS logic gates are built up using pairs of n- and p-channel metal oxide semiconductor field effect transistors (MOSFETs). Fabrication usually starts with an n-type substrate into which is diffused a p-type well. The source and drain diffusions of the n-channel MOSFET are made into the p-well, those of the p-channel MOSFET are made directly into the n-type substrate. Figure 2.1 shows a cross-section through a typical CMOS pair, in this case wired to form an inverter.

The device used for the initial study to determine that the principles and methodology of the investigation were appropriate was the dual D-type flip-flop type 4013. The fact that this is a dual circuit was useful since the basic chip, and hence the processing parameters, were common but only half of the whole chip had to be simulated. The circuits of the first batch to be tested were encapsualted in plastic and stressed at a fairly high temperature. This batch was intended for a rapid evaluation of the size of the changes of the test parameters during the stress and the statistical lifetime of the devices. In fact, the devices proved to be very resistant to change so that more study was expended on them than initially expected. All the other devices were encapsulated in ceramic packages.

The 8-bit shift register represented a higher level of integration but it had the advantage that the basic element was repeated so that simulation was not too involved.

To obtain specific information about the design, layout and device and processing parameters for commercial devices is not simple because of commercial secrecy. This resulted in some effort being expended in reverse engineering the devices used. For the next more complex circuit in the sequence, therefore, a semi-custom test chip was used. The manufacturer supplied the SPICE parameters in commercial confidence and the logical design was under experimental control. The requirement of this circuit was to provide a vehicle for the next stage of complexity so that the tests that had been found useful in the previous, smaller circuits could be implemented in a manner that was close to that which would be used in an industrial environment. In order that the degree of complexity did not increase too rapidly the design was implemented in two sections. In the first type of circuit there was a simple 4 × 4-bit multiplier, and in the second the outputs from the multiplier were connected directly to the parallel inputs of an 8-bit shift register.

This design enabled confirmation of the previous work using the 4014 shift register but also introduced a section with combinational logic. This has rather different test requirements from sequential logic. The results on this circuit are not complete and are only reported here in outline.

Figures 2.2 (a), (b) and (c) show gate level diagrams of the 4013, 4014 and of the semi-custom (multiplier) chips used in these studies. The transistor level circuits of each of the gates used in the various devices are shown in Figure 2.3. Table 2.1 shows the basic information on the devices used.

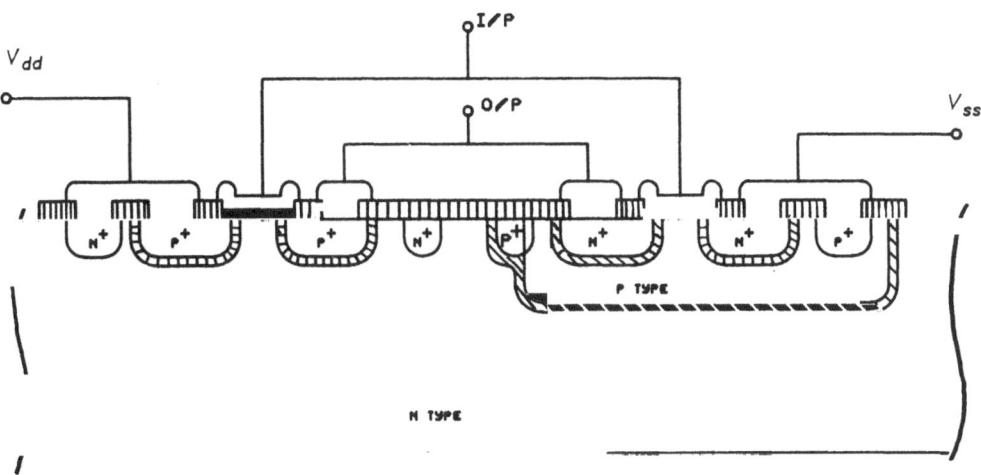

Fig. 2.1. Cross-section through a typical CMOS inverter showing the diffusions and the depletion regions.

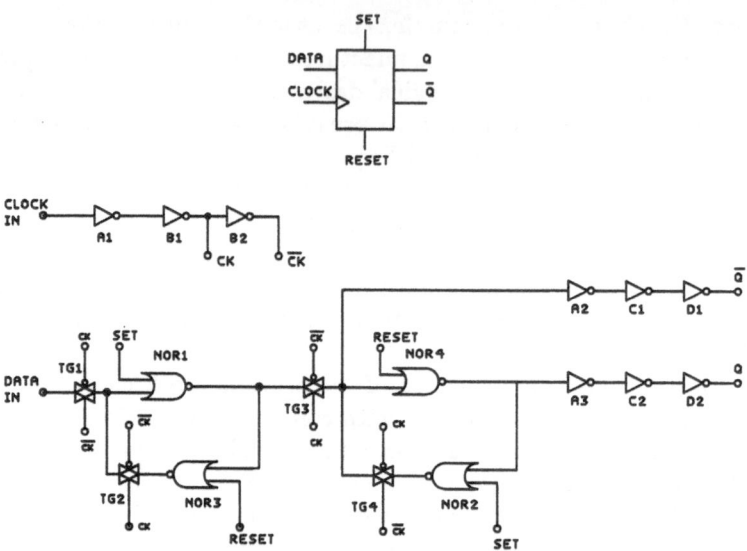

Fig. 2.2(a). The gate-level diagram of half a 4013 dual D-type flip-flop. TG1–TG4 are bidirectional transmission gates. NOR1–NOR4 are NOR gates. Inverters are labelled A_n, B_n, C_n and D_n indicating different gate dimensions.

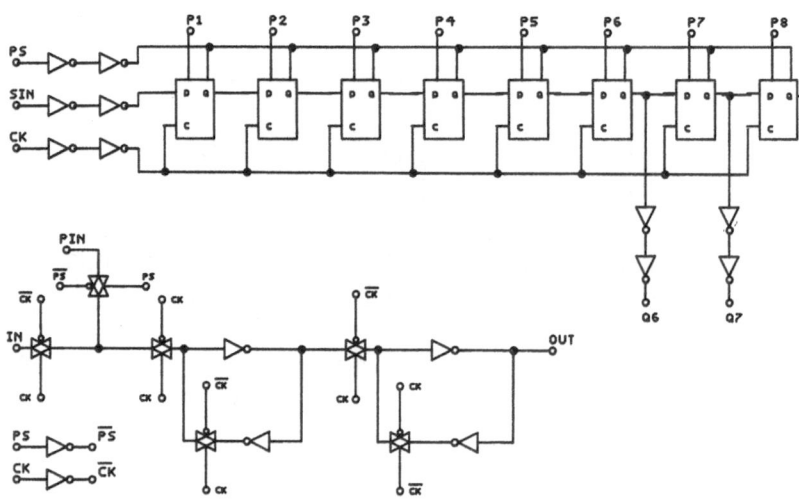

Fig. 2.2(b). The 4014 8-bit shift register. Each stage consists of a D-type flip-flop. A gate-level diagram of one stage is shown.

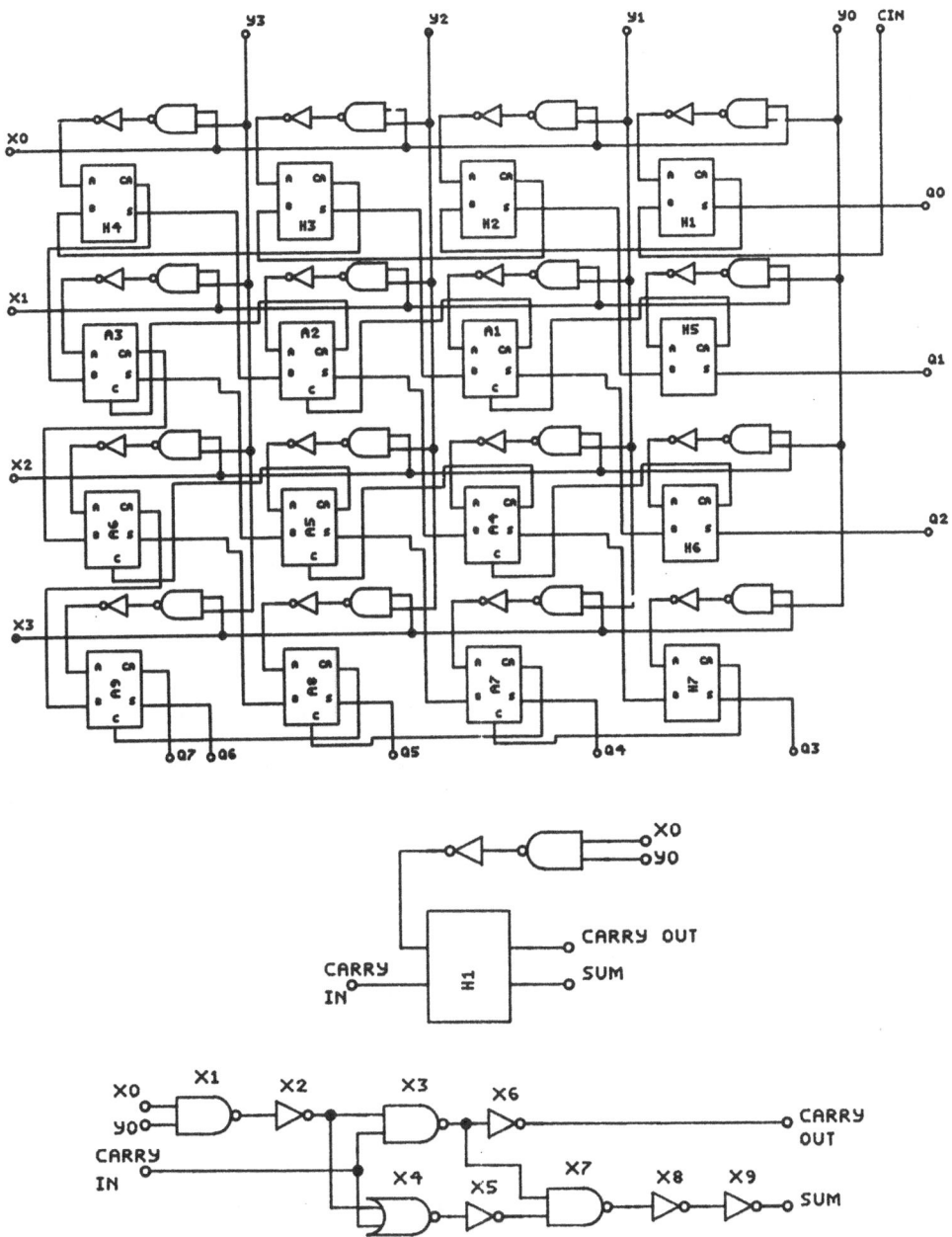

Fig. 2.2(c). The semi-custom multiplier chip. Also shown is a gate-level diagram of
the half-adder used in the simulation work.

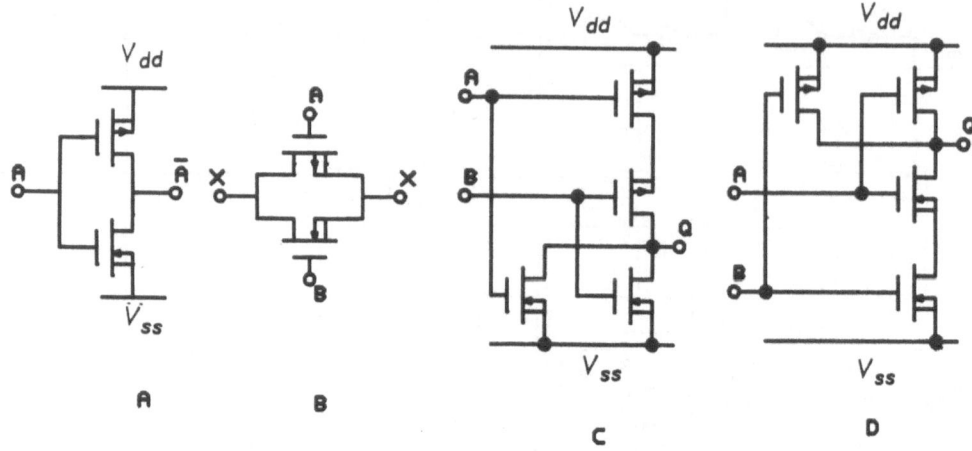

Fig. 2.3.   Transistor-level diagram of (A) CMOS inverter, (B) bidirectional transmission gate, (C) NOR gate and (D) NAND gate.

Table 2.1. Information on the Devices used in this Study

Batch no.	Device type	Description	Encapsulation	Manufacturer	Stress
1	4013	Dual D- type flip-flop	Plastic	National Semiconductor	Temp. (182°C)
2	4013	Dual D- type flip-flop	Ceramic	Motorola	Temp. (220°C)
3	4014	8-Bit shift-register	Ceramic	Motorola	Radiation
4	4014	8-Bit shift-register	Ceramic	Motorola	Temp. (220°C)
5		4 × 4 Multiplier	Ceramic	MCE	
6		4 × 4 Multiplier + 8-bit shift-register	Ceramic	MCE	

2.3. SPICE CIRCUIT FILES AND DEVICE PARAMETERS

The integrated circuits studied were simulated using a standard package, PSPICE.

Although PSPICE runs more slowly than SPICE, and is more restrictive in the size of circuits that can be simulated, it has a major advantage in the inclusion of a graphics postprocessor. This postprocessor allows the output of the simulation program to be displayed clearly, with multiple traces on the same graph for ease of comparison.

Before any circuit can be simulated it first has to be 'described' in a standard SPICE input file. The circuit is described by attaching each circuit element to nodes which are designated by a node number. These node numbers are used to indicate the connections within a circuit.

A sensible method of describing digital circuits is to set up each type of logic gate used in the device as a subcircuit. The subcircuits can then be connected together to form the final circuit. Since one type of gate may be used very frequently in a digital circuit, the advantage of this method soon becomes obvious.

As well as describing the circuit it is necessary to define the supply voltage and the shape of any input waveforms. The type of analysis to be performed by SPICE also needs to be specified. A number of analysis 'types' are available:

DC Sweep —A voltage or current source is swept through a range of values and the bias point of the circuit is caculated for each value.

Bias Point —Calculates the initial DC bias point of a circuit.

Transfer Function —Calculates the small signal gain, input resistance and output resistance of a circuit.

DC Sensitivity —Calculates the sensitivity of one node voltage to each device parameter.

AC Analysis —Calculates the small-signal frequency response of a circuit over a specified frequency range.

Noise —Calculates the noise contribution from each device and does an RMS sum of one output node.

Transient Response —Calculates the response of a circuit from time zero to a specified time.

The transient analysis is most useful for the kind of work performed here. The propagation of a waveform can be observed through a circuit, delays measured and the effect of changes to circuit parameters noted. The bias point of the circuit is automatically calculated before the transient analysis is performed.

The SPICE program requires a transistor-level description of the circuit of the device to be simulated. This information can usually be obtained from the manufacturer, or from data books. In some cases, however, it may be necessary to use scanning electron microscope (SEM) pictures to determine details of the transistor-level circuit.

For an accurate simulation it is of vital importance that the modeling parameters used for the active devices, in this case n- and p-channel MOSFETs, are as close to reality as possible. Obtaining detailed information on commercially available devices can be difficult as manufacturers tend to be unwilling to divulge information.

In the case of the semi-custom test chips used in this study, a full transistor-level description of the gates and a set of SPICE parameters were supplied by the manufacturer. For the other devices used only a limited amount of data was supplied, usually little more than the channel width/length ratio for the various transistors in the circuit.

To enable the simulation of a CMOS circuit to be performed at all, certain modeling parameters are essential. As well as the channel width and length of the transistors in the circuit, it is necessary to know the threshold voltage (V_{Th}) and the gain factor $(K_n$ or $K_p)$ of both the n- and p-channel MOSFETs.

SPICE calculates the drain current of a MOSFET according to the following equations:

In saturation

$$I_{drain} = \frac{K_p}{2} \frac{W}{L} (1 + \lambda V_{ds})(V_{gs} - V_{Th})^2.$$

In the linear region

$$I_{drain} = \frac{K_p}{2} \frac{W}{L} (1 + \lambda V_{ds}) V_{ds} [2(V_{gs} - V_{Th}) - V_{ds}],$$

where W is the channel width, L is the channel length and K_p is the gain factor of the transistor (for hole carriers) defined thus:

$$K_p = \frac{\mu_o \epsilon_{ox}}{t_{ox}}.$$

Here λ is the channel length modulation factor, V_{ds} is the drain-source voltage, V_{gs} is the gate-source voltage, V_{Th} is the threshold voltage at any bias condition defined by

$$V_{Th} = V_{To} + \gamma \left(\sqrt{2\phi - V_{bs}} - \sqrt{2\phi} \right),$$

ϵ_{ox} is the oxide permittivity, μ_o is the surface mobility, t_{ox} is the oxide thickness, γ is the bulk threshold parameter, ϕ is the surface potential and V_{bs} is the bulk-source voltage. If a value of γ is not known a default value of zero is used; V_{Th} is then equal to V_{To}.

Values of V_{Th} and K_n or K_p can be estimated for some of the older device types (such as the 4000 family of CMOS devices) from a knowledge of the technology used in their fabrication. However, it is also possible to measure these parameters for the input transistor pairs of the actual devices and then use this information for all the devices of the circuit either directly for V_{Th} or by scaling with the geometries for the gain factors. The geometries can be obtained from images of the metallization obtained by optical or SEM microscopy.

Since for the 4013 and 4014 devices no information was known about the value of the threshold voltage and gain factor, the threshold voltage and gain factor of the transistors in the input stages of a circuit were obtained by ramping the input voltage up from $0\ V$ to V_{dd} and measuring the supply current. When the input is at $0\ V$ the p-channel transistor is switched on and the n-channel is off. As the input voltage increases past the n-channel transistor's threshold voltage the supply current starts to increase. Eventually the p-channel transistor begins to turn off and the supply current decreases again. The dominating transistor is in current saturation so that the equation given earlier holds. If the square root of this supply current is plotted against the input voltage, the slopes of the resulting curve give values for the gain factors of both transistors and the intercepts give the values for the V_{Thn} and V_{Thp} This procedure was simulated on the 4013 and 4014 devices used in this study. The results of the simulation were compared with the experimental results of the threshold voltage test (ThVT) described in Section 3.3.6 in order to confirm that the modeling parameters used in the simulation were correct.

A more serious problem occurs when the doping densities and diffusion areas are not known. Without this information the SPICE model cannot make allowance for capacitance effects. This is a serious drawback as the capacitance of the circuit greatly affects the delay and, therefore, the cut-off frequency of the device. To circumvent this problem a process of reverse engineering was applied for some of the devices. If the doping densities and diffusion areas are not supplied, SPICE uses the default values of zero for these parameters. No account will, therefore, be taken of capacitance effects. To compensate for this a capacitor can be added between the gate and common nodes of each MOSFET. This capacitor, called a nodal capacitor, should also allow for stray and track capacitances. The initial value of this capacitor can be estimated using the channel W/L ratio, and the track lengths between transisitors as a guide.

The value of the nodal capacitor used in the model then has to be checked to see if it is reasonable. To test the capacitor value it is necessary to compare the modeling results with some experimental results. The nodal capacitance can then be adjusted until the simulation results agree with the experimental results.

For the 4013 and 4014 devices studied in this work the cut-off frequency, or maximum frequency of operation as a function of supply voltage, was known from experimental measurements on the devices. The cut-off frequency was, therefore, simulated at one supply voltage using the method described later. The result of this simulation was compared with the experimental result for that device at that supply voltage. The value of the nodal capacitance was then altered and the simulation run again. This process was repeated until the simulated cut-off frequency was within 10% of the experimental result at that supply voltage. Using the values of nodal capacitance obtained by this method, the cut-off frequency at two more supply voltages were simulated. These simulation results were then compared with the experimental cut-off frequency against supply voltage curve. The results are shown in Figure 2.4. The agreement observed between the experimental and simulated results indicated that a reasonable model had been obtained.

The circuit files resulting from the above procedures for the devices studied in this work are listed in Appendix 1. In the case of the semi-custom test chip, the transistor modeling parameters have been placed in a separate library file in order to preserve confidentiality. Due to limitations on computer memory and processor speed it is not usually feasible to simulate large ICs in their entirety. Of the devices studied here, it was possible to simulate one entire D-type flip-flop of the dual 4013. For the larger 4014 and semi-custom chip it was only possible to simulate small parts of the overall circuit. The 4014 shift-register was broken down into its individual stages. Initially only a single stage was simulated. More stages were eventually added to see how a waveform propagated through the circuit.

The multiplier on the semi-custom chip presented a larger problem. The circuit was very much larger and more complex. There are few sections in a multiplier which operate without interacting with some other part of the circuit.

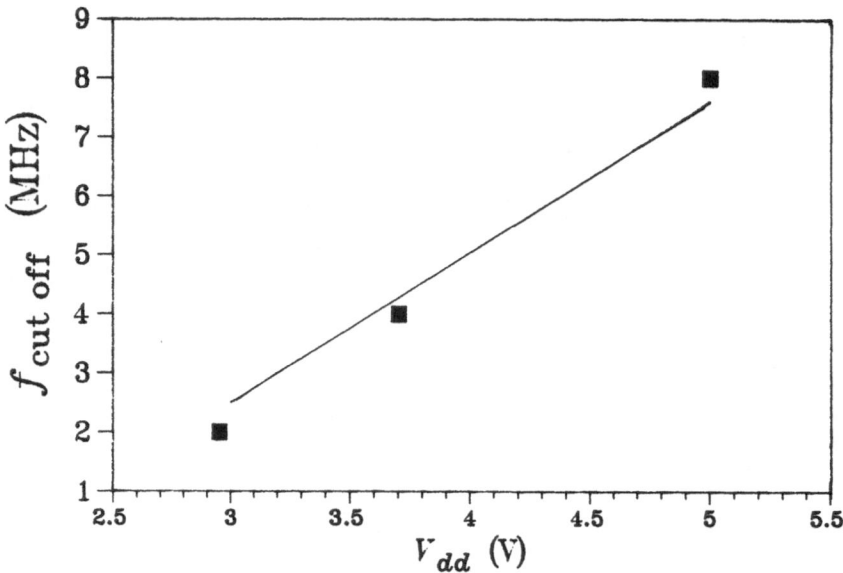

Fig. 2.4.   The variation of cut-off frequency with supply voltage for the 4013 flip-flop. The solid line shows experimental results and the squares show simulated cut-off frequencies.

The area eventually chosen for simulating was one of the half-adders [HI in Figure 2.2(c)] which is the one responsible for operating on the least significant bits of both the multiplier and the multiplicand. This section was considered to be the best part to simulate for two reasons. Firstly, the half-adder is not too large a circuit element. Secondly, it can operate without affecting any other part of the multiplier circuit. As this section of the multiplier can also be tested experimentally by itself it was possible to relate the simulation directly to the experimental results.

Before proceeding to a discussion of simulation techniques a problem should be mentioned. The transmission gates in devices such as the 4013 and 4014 can cause convergance problems in SPICE. The main problem is in the calculation of the initial bias point. It was found that by specifying the initial voltages each side of a transmission gate these initial problems could be resolved.

2.4. SIMULATION PROCEDURES

The 4013 was simulated in its entirety. For the larger chips it was only practical to simulate sections of the device. When only part of a device can be simulated the results cannot be used in fault location diagnostics. However, the results of the simulation are still useful in a sensitivity analysis.

In this study a variety of simulation procedures were used, varying from device to device. These procedures will now be described and an indication as to why each test was considered useful will be given.

The cut-off frequency of the 4013 was measured experimentally by toggling the clock and data inputs, with the data at half the clock frequency, and observing the output waveform. As the frequency of the inputs is increased towards failure the originally square output waveform becomes rounded. Eventually the peak-to-peak amplitude begins to drop, since the transistors do not have time to fully charge and discharge the nodal capacitances. The circuit was classified as having failed when the amplitude of the output waveform reached half the supply voltage.

A different method was used to find the cut-off frequency of the 4014 shift register. This was tested using HP8080 and HP8182 automatic test equipment. A data stream was clocked through the shift register at increasing frequencies until the output waveform no longer agreed with the expected output waveform.

Simulating the cut-off frequency of the sequential 4013 and 4014s was rather a long process. The output and clock waveforms had to be defined and SPICE run. Based on the results of this simulation run the frequency of the input and clock waveforms was increased or decreased and SPICE run again. This process was repeated many times until the simulated output of the device reached the failure criterion. This failure criterion was based on the experimental results. As the clock frequency increased towards the cut-off frequency the originally square output pulses became rounded. Eventually the amplitude of the output pulses began to decrease. When the amplitude of the output pulse reached $V_{dd}/2$ the device was said to have failed, and the clock frequency was taken as the cut-off frequency.

Simulating the cut-off frequency has the advantage that the result is a single number which can easily be compared with experimental results. The disadvantage is the length of time the simulation takes to perform.

The cut-off frequency is a convenient method for examining the delay in a circuit. An alternative method for CMOS devices, however, is to look at the supply current waveform. CMOS devices draw very little current unless some of the transistors in the circuit are switching. When switching does occur there is a brief pulse of current. If all of the input vectors are activated in turn a current signature is observed. This current signature consists of a series of peaks of varying size and shape, depending upon which transistors are switching. In this work the current signature of a device is called the transient current pulse, abbrieviated to TrCP, as part of the transient current test (TrCT).

Experimentally the TrCP is observed by toggling the clock and data inputs and monitoring the voltage drop across a resistor on the power supply line using a fast-storage oscilloscope.

The TrCP contains information on the switching time of transistors in various parts of the circuit. The switching time of the transistor can be related to the circuit delay. Simulating the TrCP has the advantage over simulating cut-off frequency in that SPICE only needs to be run once. The disadvantage of simulating the TrCP is that the result is more difficult to interpret. A single number cannot easily be obtained for comparison with experimental results.

Here the TrCP was simulated by setting up the input and clock waveforms and running SPICE. The simulated supply current was plotted out using the graphics postprocessor. The individual peak shapes and peak widths were then measured from the graph.

In the case of combinational circuits, such as the semi-custom multiplier chip, the circuit delay can be measured directly using HP8080 and HP8182 automatic test equipment.

To obtain consistent results the delay has to be measured in the same manner each time. Here the delay was measured from the time the input signal crossed $V_{dd}/2$ to the time the corresponding output signal crossed $V_{dd}/2$. The simulation was performed by simultaneously toggling the two inputs to the half-adder, with the carry-input at $0\ V$, which caused the output to switch. The input and output waveforms were then plotted on the same graph, using the graphics postprocessor, and the delay measured. The results of these simulations were then compared with the experimental results.

In the case of the semi-custom multiplier chip the delay was simulated over a range of supply voltages. The simulation results were then compared with experimental results, as shown in Figure 2.5. Even when using the modeling parameters supplied by the manufacturer the simulation results do not correspond exactly with experimental results. This indicates that care should always be exercised when drawing conclusions from simulation studies. For the 4013 and 4014 the simulated cut-off frequency at various supply voltages was compared with the experimental cut-off frequency against supply voltage curve.

When a satisfactory model has been obtained it can be used to test the sensitivity of a circuit to changes in the modeling parameters. In the case of the 4013 and 4014s the sensitivity analysis was performed using the TrCT. For the semi-custom multiplier chip the sensitivity analysis was performed using the time delay. The cut-off frequency simulation was not used in the case of the 4013 and 4014s due to the length of simulation time required to find each value of cut-off frequency.

The modeling parameters to be varied were chosen by considering which parameters were likely to change in real-life degradation. The variables chosen for examination were the threshold voltage, leakage current and mobility (represented by the gain factor K_n or K_p). As well as varying these parameters, an attempt was made to simulate the effect of track narrowing caused by electromigration by inserting a resistor into the circuit model between the output from one transistor and the input to another.

The effects of the parameter changes were simulated at various points in the circuit to test the effects of parameter changes in individual transistors, and over the whole circuit to simulate the effect of an error during the processing of the device. To change the parameters of a transistor in only one part of the circuit requires a separate gate subcircuit to be set up using a transitor with the altered parameter. The normal gate was then replaced in the model with the altered gate.

Analysis of the TrCP proved rather complex. Different peaks in the TrCP changed in shape and size depending upon where in the circuit the altered gate had been placed. Usually the largest effects were observed in the peaks associated with the switching of large output buffers. Most of the results of the sensitivity analysis discussed in the next section were taken by measuring changes in these output switching peaks.

A major problem was observed during the simulation of the delay in the semi-custom multiplier chip. In this device the outputs from the internal multiplier circuit are taken directly to large output buffers. The small internal transistors on the chip found it difficult to drive the large gate capacitance of these output buffers. A large delay was, therefore, introduced into the circuit. Simulation showed that at times the output buffer had finished switching long before the preceding internal stages. This poses an interesting question: Should the circuit delay be measured with reference to the fast output buffer or with reference to the slower internal stage? In this work the delay was in fact measured from the time when the input signal passed $V_{dd}/2$ to when the output from the large buffer passed $V_{dd}/2$, in order to be able to compare the simulation results with the experimental results.

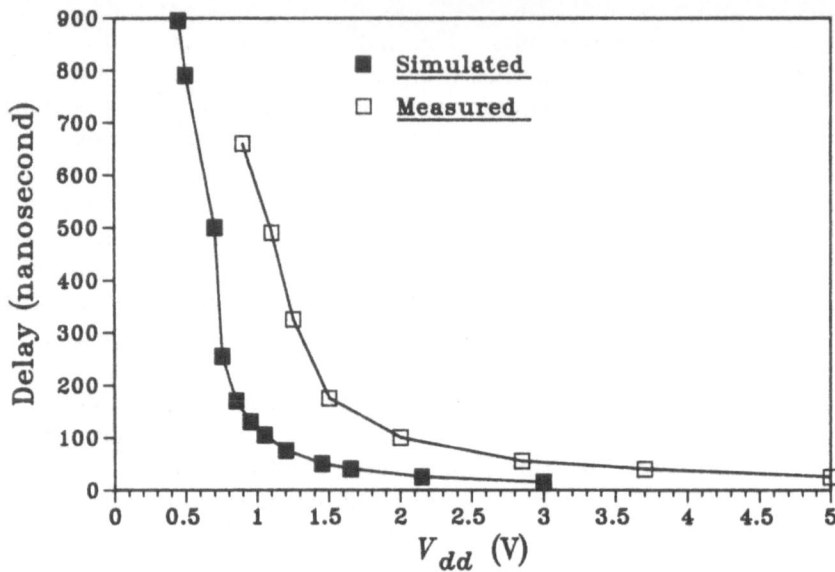

Fig. 2.5. Delay against signal and supply voltage for the semi-custom multiplier chip.

2.5. RESULTS

2.5.1. 4013 Results

The sensitivity analysis on the 4013 was performed by simulating the TrCP. The clock and data inputs were toggled, with the data input at half the frequency of the clock input, and the resulting TrCP was examined using the graphics postprocessor. Figure 2.6 shows the clock and data waveforms and the resulting TrCP. The set and reset inputs were both set to 0 V.

The larger peaks in the TrCP are produced when the large output buffers switch (these will be referred to as the output switching peaks). The small peaks are produced when only the internal transistors switch. Using the graphics postprocessor any of the peaks in the TrCP can be expanded and examined more closely.

The sensitivity of the circuit to changes in leakage current was examined by changing the value of I_S (the bulk to junction saturation current) in the transistor model. This effectively changes the bulk leakage current I_B.

The default value of I_S is 10^{-14} A. Changing this value of I_S to 10^{-8} A had no noticeable effect on the TrCP. When I_S was set to 10^{-6} A there was a very slight (26 μA) shift in the baseline of the TrCP. However, increasing I_S to 10^{-4} A had a significant effect. The baseline of the TrCP shifted up to an average value of about 2.2 mA, the level of the baseline varying according to the input conditions (see Figure 2.7). This result is interesting since the same effect was observed experimentally during the radiation stress, described later in Section 4.1.7.

The effect on the TrCP of altering the gain factor (K_n or K_p) and threshold voltage (V_{Th}) depended upon the position in the circuit of the altered transistor. The n- and p-channel threshold voltage of transistors in the master flip-flop, slave flip-flop, clock and output buffers were altered in turn. The effect on the TrCP of each change was noted. Different peaks in the TrCP were affected depending upon which transistors were altered. In all cases the output switching peaks show the greatest effect.

The width of the peaks did not increase linearly with V_{Th}. Initially the peak width increases slowly. As the change in V_{Th} becomes 'larger', however, the peak width increases more rapidly. The magnitude of parameter change needed before the peak width starts to increase rapidly depends upon the position in the circuit of the transistor which is being altered. Figure 2.8(a) shows a typical graph of change in peak width against change in V_{Th} expressed as a percentage of the original values. A comparison was made between the simulated TrCP produced when V_{Th} of the n-channel transistor in each section of the circuit was similarly increased. The results for an increase in V_{Th} of 0.8 V are shown in Table 2.2. These results indicate that the circuit is far more sensitive to change near the output stages, in the slave flip-flop or in the output buffers, than to changes nearer the inputs, such as the master flip-flop or the clock buffers.

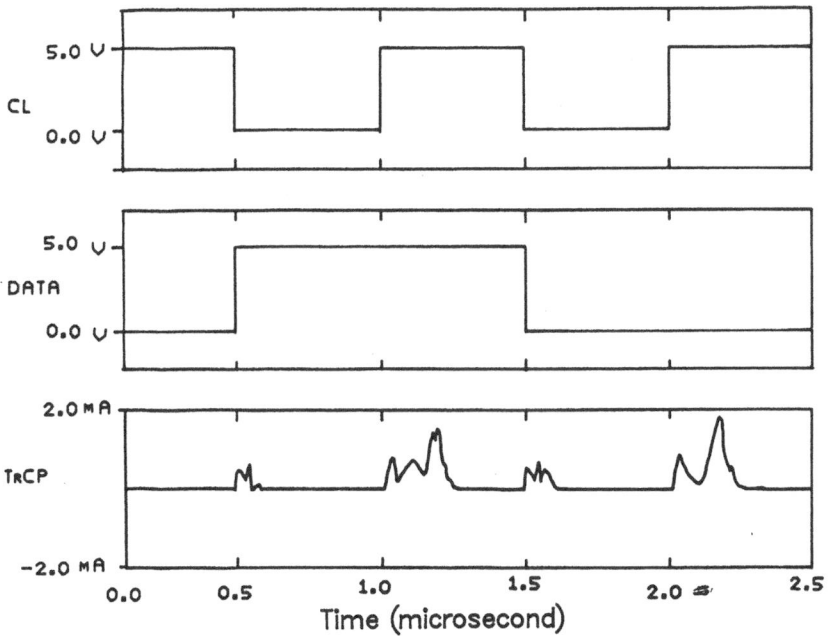

Fig. 2.6.   The simulated TrCP for the 4013. Also shown are the clock and data waveforms.

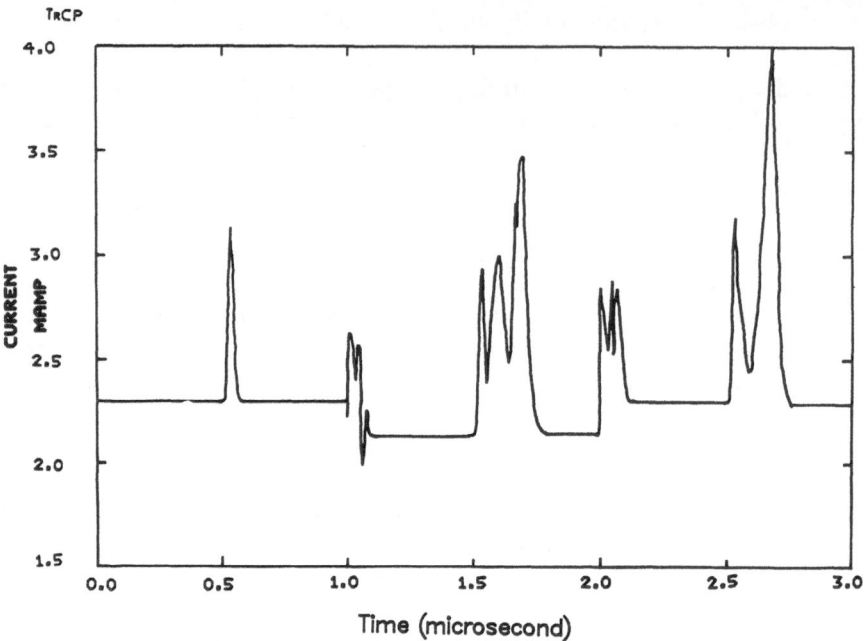

Fig. 2.7. Simulated TrCP for the 4013 D-type flip-flop with a large increase in the leakage current, I_s.

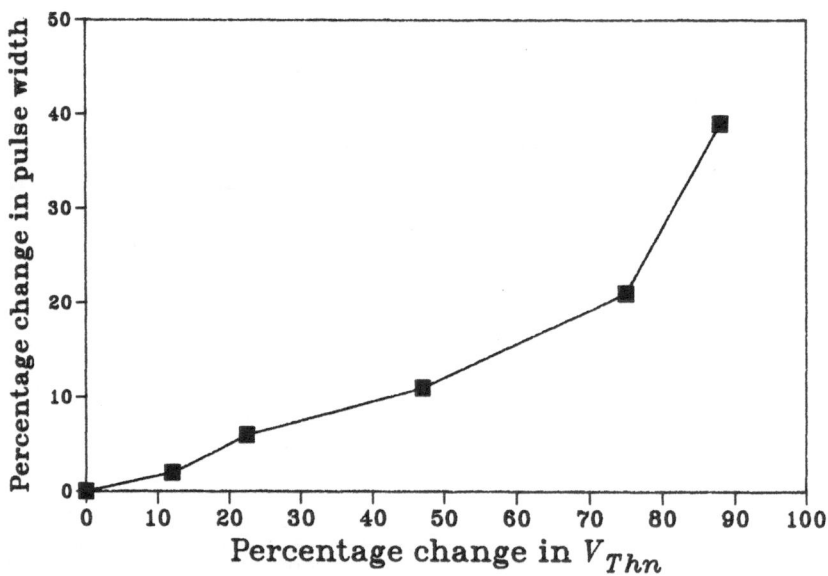

Fig. 2.8(a). The simulated change in width of one of the output switching peaks of the TrCP for the 4013 flip-flop against the change in threshold voltage of the n-type MOSFET in the inverter C1[see Figure 2.2(a)].

Table 2.2. Effect of a 0.8 V Increase in V_{Thn} of an n-Channel Transistor
Throughout the 4013 Circuit

Section	Change in peak width (nanosecond)
Master flip-flop	5.4
Slave flip-flop	24
Clock buffers	9.6
First output buffer stage	40
Second output buffer stage	13.6

Changing V_{Th} of all the transistors in the circuit tends to have little effect. Both n- and p-channel threshold voltages tend to shift in the same direction under stress. A positive shift in V_{Th} of the n-channel transistor would be compensated for by the reduction in the (negative) V_{Th} of the p-channel transistor.

Similar results to the above were observed for changes in the gain factor K, although the circuit appeared to be more tolerant to small changes in K than to changes in V_{Th}. Figure 2.8(b) shows a graph of change in peak width against change in K_n expressed as a percentage of the original value. After an initial small change the peak width begins to change rapidly. Figures 2.8(a) and (b) both show the result of changes to the n-channel transistor in inverter Cl [see Figure 2.2(a)]. Reducing the gain factor of all of the transistors in the circuit by 11% caused an overall increase in delay through the circuit. The width of the peaks in the TrCP increased by about 10%.

2.5.2. 4014 Results

The 4014 shift register can be broken down into its individual stages for the purpose of simulation, each stage consisting of a D-type flip-flop. For every stage there is a clock buffer and a buffer of the parallel/serial (PS) line. Simulation of the 4014 was firstly performed on a single stage of the shift register. All of the clock buffers and parallel/serial buffers were included to keep the loading on the clock and PS lines correct. The delay through this stage of the circuit was then calculated. By adding more stages, including some of the output stages, into the simulation it was possible to observe which part of the circuit was responsible for the delay.

There was no noticeable change in delay when the number of internal stages simulated was increased. When an unloaded output stage was added, however, an increase in delay was observed. These results indicate that the delay in the circuit is due to the clock and output buffers.

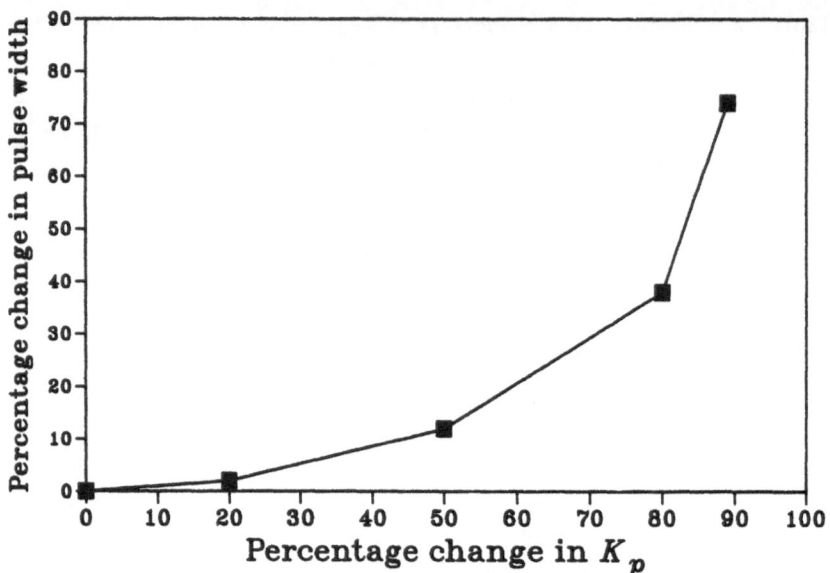

Fig. 2.8(b).   The simulated change in pulse width of one of the output switching peaks of the TrCP for the 4013 flip-flop against the change in the gain factor K_n of the n-type transistor of the inverter C1 [see Figure 2.2(a)].

A sensitivity analysis similar to that performed on the 4013 was also performed on the 4014. Figure 2.8(c) shows the simulated change in pulse width of one of the output switching peaks of the TrCP against change in V_{Th}. A comparison between Figures 2.8(c) and 2.8(a), the identical results for the 4013, shows that both devices react in a similar manner to changes of the parameters, although the 4014 seems to be slightly more sensitive than the 4013.

2.5.3. Semi-Custom Multiplier Results

Because a complete list of SPICE parameters was available for the semi-custom multiplier chips it was possible to perform a more detailed sensitivity analysis than for the previous test chips. The effect of changes in V_{Th} and leakage current were simulated as before. Instead of just altering K_p, however, the effects of changes in t_{ox} and μ_o (where $K_p = \mu_o \epsilon_{ox}/t_{ox}$) were simulated. For additional information the effects of altering the substrate doping (N_{sub}) and the effect of track narrowing were also simulated.

The effect of changing V_{Th} of the transistors in the circuit was simulated first. The effects caused by changes to V_{Th} in the individual inverters X2, X5 and X8 [see Figure 2.2(c)] were then tested. In each case V_{Th} was changed by 0.4 V and the effect on the circuit delay measured.

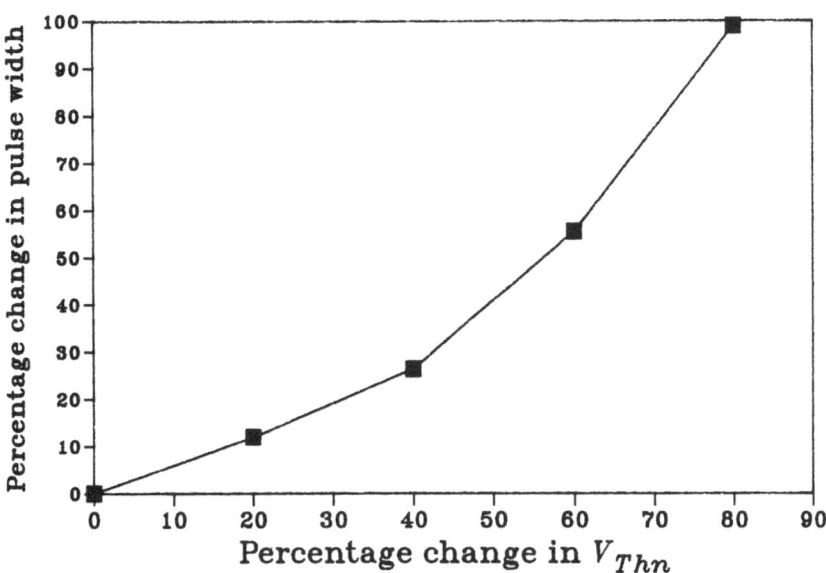

Fig. 2.8(c).   The simulated change in width of one of the output switching peaks of the TrCP for the 4014 shift register against the change in the threshold voltage of the n-type MOSFET in one of the output buffers.

The simulation was run with supply voltages of 3.0 and 1.3 V. The higher supply voltage allowed the sensitivity at normal operating voltages to be gauged. The lower voltage is just above the sum of the threshold voltages of the n- and p-channel transistors; it is the lowest voltage at which the circuit would be expected to operate.

The results of the simulations discussed above are given in Table 2.3. At a supply voltage of 3.0 V there is little change in delay regardless of where in the circuit V_{Th} is altered. At a supply voltage of 1.3 V, however, the delay changes markedly, especially for changes in inverter X8.

Figure 2.9 shows the variation of delay with change in V_{Th} for the inverters X2 and X8. Also shown is the variation of delay with V_{Th} at an elevated temperature $(60^0 C)$ for inverter X8. The sensitivity does not appear to be affected by this increase in temperature. All of the above curves were simulated using a supply voltage of 1.3 V.

The simulation was run with the oxide thickness increased by 10, 20 and then 30% with the supply voltage at 1.3 V. The effects of these changes are shown in Table 2.4. The delay decreases slightly with increased oxide thickness, the result of two conflicting processes. The gain factor is reduced by the increase in t_{ox} and the oxide capacitance of the transistors is reduced. The device tends, therefore, to be tolerant of changes in t_{ox}.

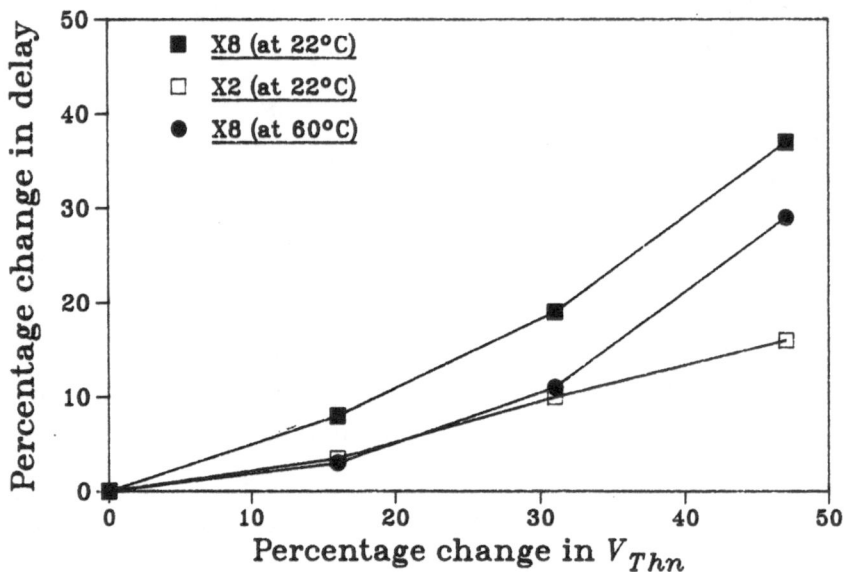

Fig. 2.9. Simulated change in the delay through the semi-cus..om multiplier chip against change in the threshold voltage of the n-type transistors in inverters X8 and X2 at a temperature of 22°C. Also shown is the change in pulse width at elevated temperature 60°C.

Table 2.3. The Effect on Delay of a 0.4 V Change in V_{2h} of Various Transistors Throughout the Circuit of the Semi-Custom Multiplier Chip

	Delay	(nanosecond)
V_{dd}	1.3 V	3.0 V
Normal circuit	83.4	20.4
All transistors	83.0	15.2
INVX2	83.0	20.2
INVX8	84.1	20.4
INVX8	90.0	20.7

Table 2.4. Effect of an Increase in Oxide Thickness, t_{ox}, on Delay at a Supply Voltage of 1.3 V for the Semi- Custom Multiplier Chip

Change in t_{ox}	Delay (nanosecond)	Change in delay(%)
0	83.4	0
10	80.2	- 3.8
20	77.9	- 6.6
30	75.0	-10.1

Table 2.5. The Effect of a Change in Mobility on Delay for a Supply Voltage of 1.3 V for the Semi-Custom Multiplier Chip

Change in mobility(%)	Delay (nanosecond)	Change in delay(%)
0	83.4	0
5	84.3	1.1
10	87.8	5.3

The effects of an overall 5 and 10% reduction in mobility are shown in Table 2.5. A 10% change in mobility (which is the same as a change in K) causes only a 5% increase in delay, compared with a 10% change in peak width in the 4013 TrCP for a similar change in gain factor. The semi-custom chip, therefore, appears to be much less sensitive to change in gain factor.

Changes in substrate doping and leakage current of up to 50% had no measurable effect on the circuit delay. These simulations were again performed at a supply voltage of 1.3 V.

The effect of track narrowing was investigated by introducing a resistor in series with the output of inverter X8. Resistor values in excess of 1 kΩ were needed before there was any measurable effect on delay. This indicates that metal tracks between transistors would need to become very narrow before they started to affect the circuit performance.

2.6. CONCLUSIONS

A series of simulations were performed on the 4013 D-type flip-flop, the 4014 shift-register, and a semi-custom multiplier chip in order to evaluate the sensitivity of the circuits to changes in the transistor characteristics. The sensitivity of the circuits was calculated for changes over the whole circuit and for localized changes throughout the circuit.

All of the devices proved to be very insensitive to changes in leakage current; increases of over three orders of magnitude had no noticable effect. Similar results were obtained for changes in the substrate doping density and for the effect of track narrowing. The devices seemed to be more sensitive to changes in threshold voltage and gain factor.

Although there was some variation between the devices in the magnitude of the changes caused by alterations in V_{Th} or K, the results were very similar. Small changes in V_{Th} or K had relatively little effect on the simulation; larger changes, of the order of 90%, had significantly more effect.

These results indicate, therefore, that CMOS is a very tolerant process. CMOS devices should perform adequately even when there are large deviations in the transistor characteristics. This means that very sensitive tests need to be devised in order to give early warning of devices departing towards failure.

The simulations indicated that the sensitivity of a test could be increased by performing the test at low signal and supply voltages. The test would be most sensitive when the device is operated at just above the sum of the n- and p-channel threshold voltages.

Simulation can, therefore, be useful for investigating how a circuit functions. It can be used to estimate what magnitude of degradation of transistor characteristics would need to take place before problems occur. The simulation could also indicate any possible design faults within the circuit in need of rectification.

REFERENCE

Syrzycki, M., 1987, *Proc. 1987 Int. Test Conf., IEEE,* 148–57.

Chapter 3

THE TESTS AND STRESS
EXPERIMENTS

3.1. INTRODUCTION TO THE TEST PROCEDURES

3.1.1. Approach to the Test Methods

As has been discussed in Chapter 1 there are, in general, two types of approaches toward reliability testing of digital ICs. Digital testing, which is the conventional approach, is normally developed on the basis of the function that the device is supposed to perform. The criterion for this approach is then function or malfunction. The problem is therefore transformed into mathematical operations which are tackled with the help of computer-aided design (CAD) techniques. This approach ignores the detail of the failure modes which involves a separate detailed analysis of the devices by experiment. The other approach is derived from the parametric characteristics such as supply current, threshold voltage, propagation delay, cut-off frequency, and so on, which are normally analog features. Instead of being functionally related and digital as in digital testing, this approach detects gradual changes or degradation of the electrical parameters which may not be large enough to cause a complete functional fault at that moment. These parameters are closely related to the chemical and physical processes of the operating devices and affect the performance of the device in one way or another. In developing test methods for this approach, the tactics involved in the development of digital testing are no longer valid since the target is now to measure gradual changes or degradation of the parameters, not necessarily the digital fault.

3.1.2. Rationale of Test Verification

Modern ICs are so reliable that a gigantic sample size has to be used to determine the degradation or failure of the device if the analysis is performed under normal operating conditions. A large sample involves a large test system and a high cost. Detailed tests would also be costly and take a long time. Even if a gigantic sample is used it still would need years of testing to allow the reliability analysis to be performed. With the development of modern technologies and techniques, the devices have become increasingly reliable such that an even larger sample size and longer testing time would probably be necessary in the future. Hence the changes or degradation of the samples needs to be speeded up, preferably without changing the failure mode from that which appears in normal use, so that a small sample size would be sufficient to perform the reliability analysis.

A normal solution to this requirement is to design accelerated stress testing, artificially 'ageing' the devices such that the gradual changes or degradation of the devices are accelerated. Results obtained by these procedures can be extrapolated to normal operating conditions so as to obtain an estimate of the device reliability.

The degradation and the failure of semiconductor ICs can normally be stimulated by various factors such as temperature, current and voltage, radiation, humidity and pressure.

In the case of temperature stress, the chemical and physical processes that are involved in many of the mechanisms that cause degradation and failure are effectively accelerated. The Arrhenius equation has proved satisfactory to describe this process:

$$R = R_o \; exp(-E_a/kT),$$

where E_a is the activation energy, which describes the energy barrier that the failure process has to overcome during the process of failing as is depicted in Figure 3.1. The activation energy in the temperature stress produces an exponential increase of the reaction rate with an increase in the temperature.

When current and voltage stress is introduced together with temperature stress, failures such as dielectric breakdown, interface charge accumulation, charge injection and corrosion are accelerated. The reaction rate of these mechanisms is given by:

$$R = f(V,T) \; f(I,T) \; R(T),$$

where $R(T)$ is again an Arrhenius function of T, and $f(V,T)$ and $f(I,T)$ describe the voltage and current stress components, respectively. Note that temperature appears in both voltage and current components.

In the case of the ionizing radiation stress, the build up of the silicon dioxide–silicon interface states and the accumulation of the oxide trapped charges cause a shift of threshold voltage. This in turn produces excess static supply current and the degradation of the mobility of channel carriers, hence affecting the performance of the device.

Despite the complex effect of the ionizing radiation species, dose rate and bias conditions on MOS devices, the simplest forms can be described as follows:

$$\Delta V_{Th} = \Delta V_{Nif_s} + \Delta V_{Notc};$$

$$\mu = \frac{\mu_o}{1 + \Delta N_{if_s} / N_o};$$

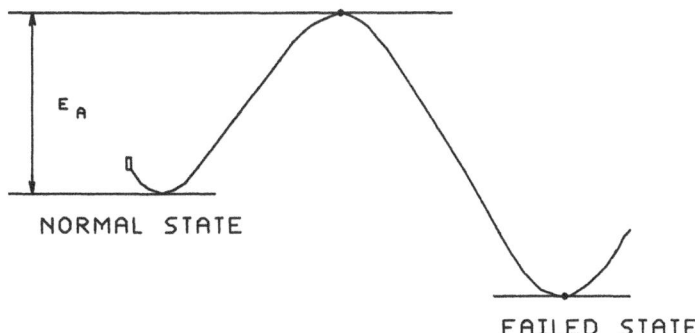

NORMAL STATE

FAILED STATE

Fig. 3.1. Schematic diagram of the energy barrier that a normal device has to overcome during the process of failing.

Table 3.1. Typical Failure Modes of MOS Technology with Relevant Accelerating Factor and Activation Energies

Failure mode	Accelerating factors	Activation energy E_a (eV)
Oxide and silicon-oxide interface surface charge accumulation	V T	1.0 ~ 1.4
Slow charge trapping	E T	1.0 ~ 1.3
Oxide shorts and breakdown	E	
Metallization corrosion	H V T	0.3 ~ 1.0
Intermetallic growth	T	0.7 ~ 1.05
MOS threshold voltage shift	V T	0.5 ~ 1.4

$$I_{ss} \sim e^{\Delta V_{Th}} \qquad \text{in the near subthreshold region;}$$

$$I_{ss} \sim f(\Delta V_{Nifs}) \qquad \text{in the far subthreshold region;}$$

where ΔV_{Nifs} is the component contributed by interface states and ΔV_{Notc} by oxide trapped charge, μ_o is the channel carrier mobility with no interface states and $\Delta N_{ifs}/N_o$ is the increase in interface states divided by a constant.

In the case of humidity (H) together with the temperature stress, the chemical and electrolytic corrosion and threshold shifts are usually accelerated.

A summary is given in Table 3.1 to show which factors contribute to the failure process and which factors may be increased to accelerate the stress and thus increase the failure rate. It can be seen from the table that the relevant factors of most of the significant failure mechanisms can be increased in one way or another and, in particular, temperature (T) electric field (E) and voltage (V) are all parameters very efficient in affecting the acceleration of the degradation and thus causing device failures.

A few more points regarding activation energy are worth mentioning. As is indicated in Table 3.1, each process should correspond to a specific range of activation energies, even though some of these ranges may overlap. However, this is not always the case since the accelerated stress often produces several processes. The condition of the accelerated stress should be carefully chosen so that the expected failure mechanisms can be efficiently accelerated and unexpected failure mechanisms will not be stimulated so much as to be dominant over the former. Then the activation energy obtained will be valid for the extrapolation of the data from the accelerated lifetime test to the life expectancy of the device in field use.

It should be possible to choose a small sample size to perform the analysis provided that the stress factors and their accelerating rates are carefully determined so that the expected failure mechanisms can be more efficiently stimulated than the unexpected ones. How small the sample size should be depends, however, on the availability and the efficiency of the experimental facilities and manpower and on what we expect to gain from the analysis. For digital testing a relatively large sample size may be used. For parametric testing a sample of 20–50 devices should be enough. This is not too small a sample on which to perform statistical analysis reliably and also is not so large as to increase the cost of the system involved drastically; hence the time and manpower needed to perform the intermediate tests and results interpretation can be kept within a reasonable range. In our experiment we used batches of 25 devices to perform the analysis.

It is also important to have the right type of devices to work with. To start with, small scale integrated (SSI) and medium scale integrated (MSI) combinational circuits are a better choice. The function performed by combinational circuits is normally derived from the states of the primary inputs of the circuits and is independent of the timing of the circuit. By primary input we mean the circuit input which can be directly accessed through the external pins. So long as no redundant circuitry exists, a detailed understanding of the internal circuitry and layout of the device becomes less crucial to testing since all the internal states are uniquely determined by going through all state vectors of the primary inputs. The derivation of the test procedures and the interpretation of the results are then much more straightforward. This greatly assists the development of the knowledge as a whole. It is inevitable, however, that sequential circuits have, sooner or later, to be included in any program that does reliability testing since VLSIs implement complicated sequential logic. For our own purpose, the CMOS D-type flip-flop is chosen for the first stage. This is a simple sequential logic circuit in which both gates and transmission gates are implemented and both combinational and sequential logic can be accomplished. Most of the internal components of the circuit can be directly accessed through the input pins. The 4014 8-bit shift register served for the second stage in our study. Basically, the 4014 is just a chain of duplicated 4013s. A study of the 4013 provides the information of direct use to the study of the 4014, and a comparison of the results from both 4013 and 4014 tests may provide the information that indicates the change of the reliablity concerns from one aspect to another in the progress of the SSI to MSI.

To utilize these findings in an industrial setting, an extension of the work from SSI and MSI to LSI and VLSI is necessary. It is not just a simple case of adopting the techniques developed for SSIs and MSIs. First of all, the large number of the components and the complexity associated with VLSIs force us to make compromises between the fault coverage and the cost of the tests. Secondly, the mechanisms of failures associated with VLSIs may differ from those of SSIs and MSIs. The techniques developed for SSIs and MSIs have to be rejustified to serve for VLSIs. In our work, the semi-custom 4×4-bit multiplier and shift-register LSIs are used. As opposed to the previous experiments, commercial circuits were not used here. This is because a detailed understanding of the circuit parameters such as process parameters, layout parameters, simulation parameters, and so on, is very important for the analysis of the results.

3.2. ACCELERATED STRESS METHODS

In the previous section a brief account of accelerated stresses has been given for thermal, electrical and ionizing radiation, and so on. In this section we present a more detailed discussion of thermal, electrical and ionizing radiation stress.

In general, stress systems consist of two basic sections. One part is to provide well controlled stress on the sample with reliable protection against runaway for both the devices being stressed and the stress system. The other part should provide a flexibility over the stress parameters and the functional and parametric monitoring of the devices being stressed.

We first of all discuss the thermal and electrical stress systems. The schematic diagram of such systems is shown in Figure 3.2. There are basically three parts: the electrical stress, the thermal stress and the controller.

Two control units are used in the system. The system controller generates the clock signal to exercise the circuit block and the logic for the time and functionality display as well as the timing signal for the detection circuit. Electrical stress on the devices being stressed is accomplished by the exercise circuit through its driver. The outputs of the devices are sampled by the sampler and the signals are compared with reference signals to check the functionality of the device, and finally the results are stored in a random access memory (RAM). The time and functionality displays provide the sample time and the functionality of the devices being stressed at that time. The stress board which accommodates the devices being stressed and the electrical circuitry resides in the temperature chamber which provides thermal stress to the devices. The data collection controllers are normally microcomputers with an input/output interface. These are only used between each stress test run to read the functionality results stored in the memory and therefore may be detachable from the system when it is not being used.

A general exercise circuit for the digital circuit should include a pattern generator and its driver and protection circuitry as is depicted in Figure 3.3. The pattern generator produces the exercise pattern for the device under test (DUT) and this can be either static or dynamic as is needed for the specific purpose. It has been generally accepted that dynamic exercise is a better choice for modern ICs. In principle, the exercise pattern of dynamic stress for combinational logic should be so designed that all the inputs go through a similar pattern with a delay between each other and have roughly the same time in both logic–0 and logic–1 states. The exercise patterns for sequential logic are, however, more complicated. Basically, to design an exercise pattern for a specific type of device we should take care to simulate the functions which the circuit normally performs in reality. For example, the exercise pattern for a memory chip should normally reveal that the read/write command is more frequently issued than the chip select command. That is, the exercise frequency of the input data, address and read/write are higher than that of chip select pin. In addition, an attempt should be made to cover as many state vectors as possible in the exercise pattern. This will ensure that the entire circuit is being stressed.

The buffers and resistors R, R_s serve as the driver and protection components, respectively. Each buffer may drive one or more circuits. Resistor R should have a value of the order of kilo ohms and R_s about the order of tens of ohms, greatly depending on the current drawn by the device.

Care is needed in the implementation of the circuit board for thermal and electrical stress. Commercial high temperature sockets are normally available in two types: non-zero insertion force and zero insertion force. It has been found that the non-zero insertion force socket itself degrades very significantly with thermal stress. The contact failure of this type of socket can be several percent

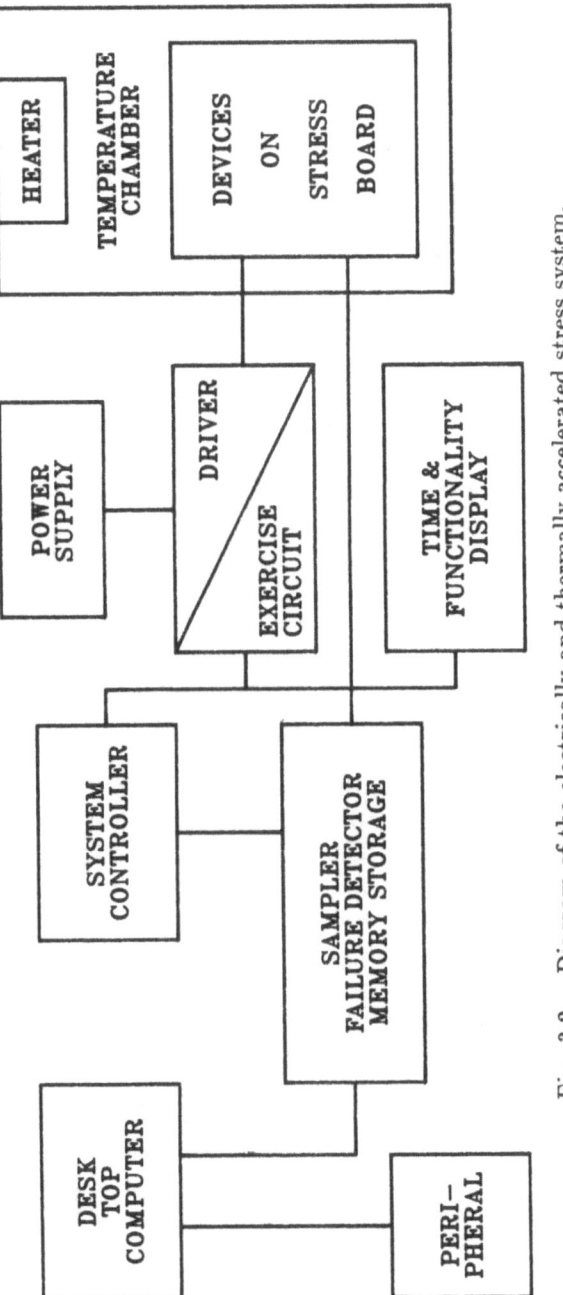

Fig. 3.2. Diagram of the electrically and thermally accelerated stress system.

43

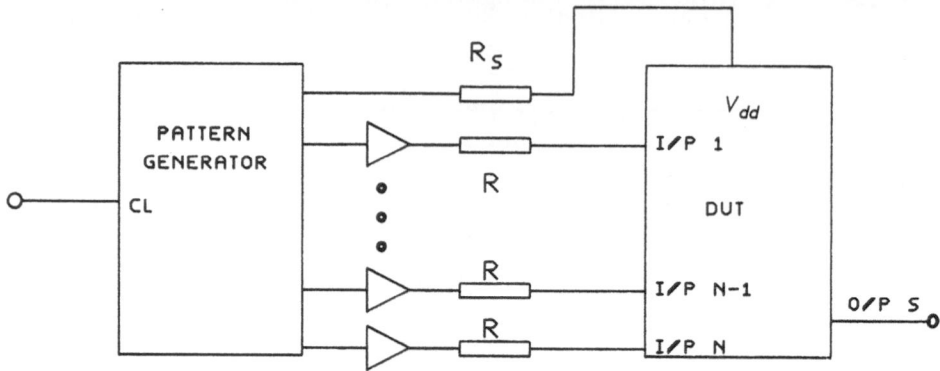

Fig. 3.3.   The general exercise circuit for the electrical stress of the digital circuit.

after a few stress-test runs. Bad contact between the device being stressed and the socket not only gives rise to great difficulty in resuming stress after each test break, but also causes the devices being stressed to degrade or to be damaged unintentionally, especially when the power supply pins are left open. The zero insertion force socket, however, does not degrade much with high temperature stress because of the mechanism it uses to make contact.

Commercial ovens can be used as temperature chambers to produce thermal stress environments. Controllers are, however, also needed. First of all, the controller of the oven should provide automatic control over the stability of the temperature in the chamber and protection against temperature runaway in case of failure of a thermocouple, fan and controller, and so on. Secondly, to prevent samples from overstress, the overshoot transient of the temperature during the oven turn-on should be minimized by adjusting the gains of temperature difference to heating power conversion if it is available and by setting the temperature about 10°C lower than the designed temperature initially and then adjusting to the precise value. A temperature controller with zero-crossing switching is recommended in order to prevent electrical spikes. It has been found that such electrical spikes can easily be picked up by the system and therefore cause extra electrical overstress on the devices being stressed and also malfunction of the controller circuitry. This is especially important for CMOS circuits since they are more vulnerable to electrical over-stress and latch-up.

The detection part circuitry may have the form as is shown in Figure 3.4. For clarity, some relevant waveforms are plotted in Figure 3.5.

44

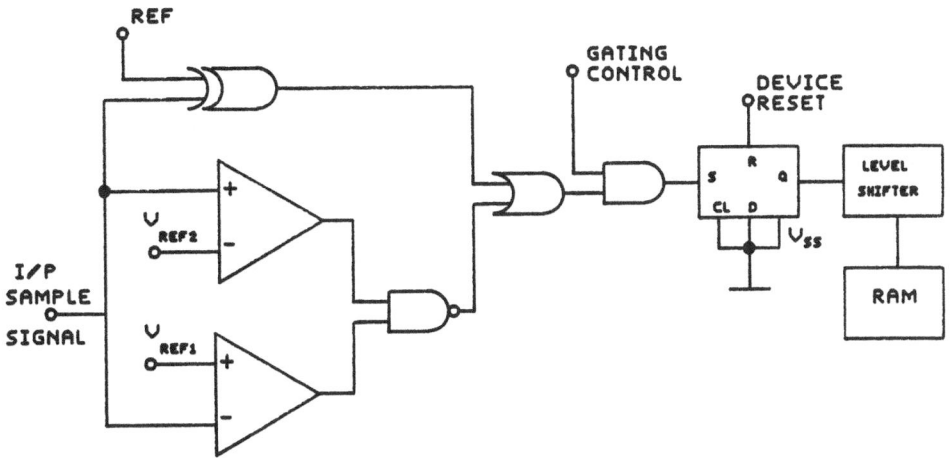

Fig. 3.4. Circuit diagram of a system that performs both logic and amplitude detection.

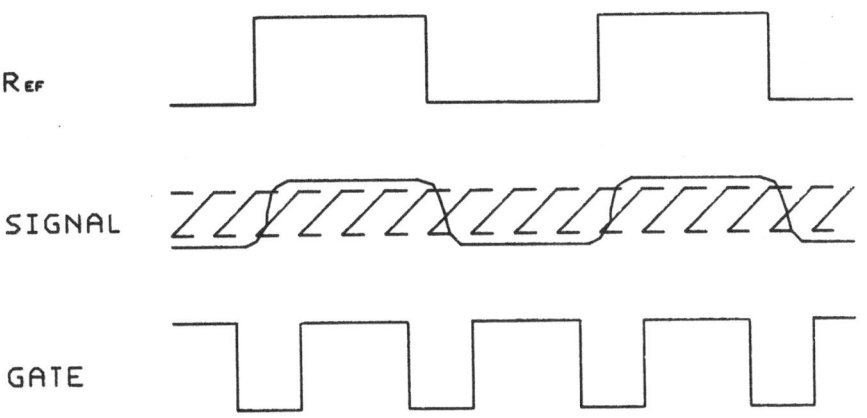

Fig. 3.5. Some relevant waveforms associated with the detection system shown in Fig. 3.4.

The output amplitude detection to test for functionality is performed by two comparators and a NAND gate. The output of the NAND gate is set only if the amplitude of the signal falls into the shaded region of Figure 3.5. The logic detection is performed by an EX-OR gate. Its ouput is set only when the signal is different from the reference. Then the two signals are OR'd and further gated before being latched into D-type flip-flop. The gating control signal shown in Figure 3.5 plays an important role in the detection circuitry since some glitches are inevitable in such systems. The delay between the reference signal and the tested signal, the relatively large loading to the devices being tested and the mismatches in the detection circuitry all produce glitches. The level shifter is used to match the stress voltage and system operating voltages.

So far we have been discussing some features of the thermal and electrical stress systems. In the following paragraphs we briefly discuss the radiation stress system we have used.

Basically, the radiation system consists of two fundamental parts: the radiation source and the bias circuit to set the devices being irradiated to a predefined state. Much work has been devoted to the study of radiation sources. The most common ones are gamma-ray and X-ray.

Comparisons between these two sources have been discussed in the proceedings of the 1986 Annual Conference on Nuclear and Space Radiation Effects. In this study we used X-ray stress because of its accessibility. The X-ray exposure was performed with a Philips PW 1009 X-ray generator. An 11 kV acceleration voltage was used with a copper anode. The characteristic K_β peak and high energy photons are removed by a 0.00254 cm thick Ni filter. The irradiation consists of the characteristic K_α peak and a continous energy range near the K_α peak. The beam current is adjusted to produce the required dose rates. The devices being radiated are all de-lidded. The bias circuit is just a normal DC bias circuit and therefore is not discussed further.

3.3. DETAILS OF THE TESTS USED

3.3.1. The Static Current Test

3.3.1.1. Introduction to the StCT

The static supply current test (StCT) measures the static supply current of the whole circuit with the device in an unexercised, static condition. There are two parts to the measurements. One part is to measure the total supply current when the device is taken through the predefined test routine, ie, a specific set and sequence of input test vectors, at a specific supply voltage. The other part is to perform $I \sim V$ measurements in a predefined voltage range, which usually follows the specified operating range of ordinary CMOS circuits (that is, between 3 and 18 volts) at selected test vectors. The schematic diagram of the test circuit

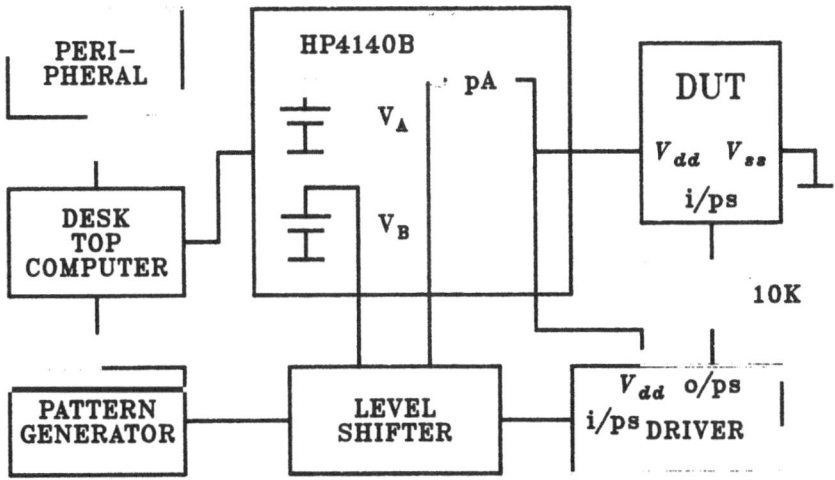

Fig. 3.6.   Block diagram of the StCT system

is illustrated in Figure 3.6. The test set is generated by a pattern generator which is controlled by the computer through the IEEE 488 parallel interface. The introduction of the level shifter provides the test circuit with the ability to perform the $I \sim V$ measurement without much cost if a variable voltage data generator is unavailable. The driver is used to supply the excess leakage current which flows from the driver to the input pin of the DUT. The HP4140B pA meter and DC voltage source is a good choice for research use since it has the very wide dynamic range needed to test several orders of magnitude variation in the static supply current despite the inherent drawback of low speed. Other faster circuits can be devised for industrial use.

The equivalent load resistance to the two power supply rails of each CMOS inverter cell, when it is not switching, is usually large enough to offer almost negligible current flow in the active components since one of the drain diodes in the pair is always in reverse bias, and the gate is isolated with an insulator. That is, the active components are not in operation when the circuit is working in the static mode. The $I \sim V$ characteristic is mainly determined by that of the internal p–n junctions subject to reverse bias and the dissipative components. Any basic cell of a CMOS gate consists of such complementary transistor pairs and these transistor pairs are connected either in series or in parallel following the rule that one is connected in series with others of the same type and the other in parallel with the others of the same type.

Therefore the total effect of the basic cell is just a number of the drain p–n junctions connected in either series or parallel and each is reverse biased.

The transmission gate is the only different type of cell apart from the basic cell mentioned above. The input and output complementary transistor pair in the transmission gate is always prevented from being connected between the two supply rails because the output of one complementary pair is connected through the transmission gate to the gate input of the following stage. The worst case is that both drain and source junctions are reverse biased when the transmission gate is biased OFF.

Therefore there is no other mechanism to generate a static supply current other than the parasitic p–n junctions subject to reverse bias by the power supply and possibly the dissipative leakages, although this is usually less competitive compared with the leakage current of the reverse biased p–n junctions in CMOS circuits.

Among the parasitic p–n junctions mentioned above, the one between the substrates of two different types of transistors is usually large in area and therefore contributes most to the static supply current in a normal CMOS circuit. The current of a reverse biased diode follows the conventional generation–recombination (g–r) current law:

$$Ig \sim \frac{A q n_i W}{\tau},\tag{3.1}$$

where A is the junction area, τ the minority carrier lifetime and W the depletion layer width which for a uniform abrupt junction is:

$$W = \left(\frac{2\epsilon_o \epsilon_s (N_a + N_d)(V_{bi} + V)}{q N_a N_d}\right)^{1/2},\tag{3.2}$$

where V_{bi} is the built-in potential of the junction and N_d and N_a are the free carrier concentrations on the n-side and the p-side, respectively. The dielectric constants ϵ_o and ϵ_s have the normal meaning. For typical CMOS technology the static supply current density has an order of about 10^{-14} A/μm^2 at 10 V supply voltage. An SSI or MSI such as a D-type flip-flop or 8-bit shift-register shows about 10^{-12}–10^{-10} A total current.

In addition, the following also contribute to the supply current: the leakage of the channel when the transistor should be OFF, and the gate and field dielectric due to surface or bulk defects and contamination by mobile charges. These leakage sources are often related to poor processing or stresses. They are, therefore, either masked by the junction leakage or neglible for a normal device.

It is clear that the normal intrinsic leakage, ie, the supply leakage current discussed above, is well defined and can be calculated for a particular circuit. Also the current of this form is independent of the states of the inputs and the internal nodes in the circuit since the current of the reverse biased p–n junction between the substrates of the two types of transistors always exists independently to the states of the active components in a normal circuit. Therefore any excess component above this intrinsic current appearing in any state of the circuit indicates a weak point or defect. Also a defect not associated with substrate p–n junctions is usually observable in the StCT only when it is excited to produce an excess supply current. Thus a defect would not be active unless the circuit is biased in some specific states where it is excited.

Now, the question is how to set the circuits to a state in which the defect is stimulated to produce the excess supply current if the site of the defect and its electrical characteristics are known. Furthermore, how can we test for all possible defects that may cause excess current in a practical size circuit? In principle we can do so for some specific types of defects if the test goes through all possible states of the circuit. But there are two problems we face by trying to do so. First, how can we develop a test set that includes all possible defects and, second, can we afford the cost of such exhaustive testing? Fortunately, these questions have been tackled partially by the people who are involved in digital testing of the ICs. We will follow in their footsteps to derive the test set for the StCT in the next subsection.

3.3.1.2. Test Pattern Generation for the StCT

The process of test pattern generation, also called the exercise pattern or the test set generation, produces an adequate test set for any particular circuit. This topic has received great attention from both the manufacturer and user of ICs, and those who work in digital testing. Some very useful strategies have been successfully developed in this area. In principle the test pattern generation methods are divided into two classes, namely those which exercise the circuit to check if it performs the specified function (referred to as function-oriented methods), and those which stimulate all faults in predefined fault lists (referred to as fault-oriented methods). Typical examples of these two classes of methods are the Boolean difference method and the sensitization method. A detailed discussion of these method is beyond the scope of this book. For completeness, however, we briefly introduce the method that follows the fault-oriented test pattern generation methodology and which was used for the derivation of the StCT.

The derivation of the test set for a given circuit in digital testing usually follows the basic steps listed below:
1. Prepare a fault list based on the knowledge of the fault types.
2. Test for a specific fault.
3. Check the fault coverage of this test vector and delete the covered ones from the list.
4. Go back to Step 2 or terminate.

Historically, the fault list is usually prepared on the basis of the stuck-at fault model despite the fact that quite a few fault types such as the bridging, the stuck-on and the stuck-open faults are generally recognized as occuring in modern ICs and are not covered by the stuck-at fault model. Obviously, the list comprises both stuck-at-0 and stuck-at-1 at all nodes of the circuit.

A specific fault in the list is then stimulated by applying a test vector to the circuit which permits the fault signal produced at the site of the fault to be propagated to the primary output of the circuit. It is obvious that this test vector also tests for some other faults included in the list, although this is unintentional when the test vector was derived. Thus all the faults that are covered by this vector should be deleted from the fault list. Finally, the decision has to be made in Step 4, whether to go back to Step 2 to test for the remaining faults in the list or to terminate the test. It seems that the loop between Steps 2 and 4 should continue until the list is emptied. However, the termination of the above loop usually occurs, for practical circuits, before the list is emptied since the economics associated with the coverage of the fault plays a crucial role in such a type of decision making. A compromise between fault coverage and the effort needed to perform it has to be made. With complex circuits it would be impractical to seek a 100% fault coverage, although this is usually achieved for the purpose of demonstration in the textbook where only gates and very simple circuits are discussed.

The procedures in the derivation of the test set in the StCT is basically the same as that of digital testing. A similar four-step process is needed. However, because of the nature of the StCT, there are two fundamental discrepancies between the two approaches. The StCT tests for latent defects instead of digital faults in digital testing. Thus instead of a fault, we refer to the list as the defect list which, of course, takes into account the digital faults that are the outcome of the defect. Secondly, in digital testing the fault is tested only if the fault signal is transmitted to the output of the circuit. But in the StCT the defect is directly observable in the supply current as long as the defect is stimulated to generate the excess current and, therefore, no consideration of transmitting a 'defect signal' to the output in needed. Bearing these features in mind, we present the procedures of deriving the test set for the StCT on basic gates by means of the sensitization method used in digital testing.

Basic cells of the CMOS circuits comprise two types of gates which we call basic Boolean function gates and transmission gates. The basic Boolean function gates can be any gates such as inverter, NAND, NOR or mixed gates. Despite the type of function a specific circuit performs, all transistors in the circuit are working under the same set of bias conditions. Each electrode of a transistor is in one of the states of negative, positive supply rail or between, depending on the way the transistors are connected and the ratio of the channel resistances of the OFF transistors across which the voltage drops occur. For example, the transistor P2 in the circuit shown in Figure 3.7 has the state table listed in Table 3.2. Here 1/2 and 2/3 indicate the voltage ratios at the nodes X and Y (which we call inaccessible nodes to distinguish them from the internal gate level nodes) when all inputs are set to logic '0' and '1', respectively provided the channel resistances of the same type of transistor are about the same. Now let us consider, within the scope of the isolated transistor, the defects in which we are interested. As examples we discuss transistors P3 and N2 of the circuit of Figure 3.7. They have the state tables listed in Table 3.3. The voltage value of the source at State 1 is

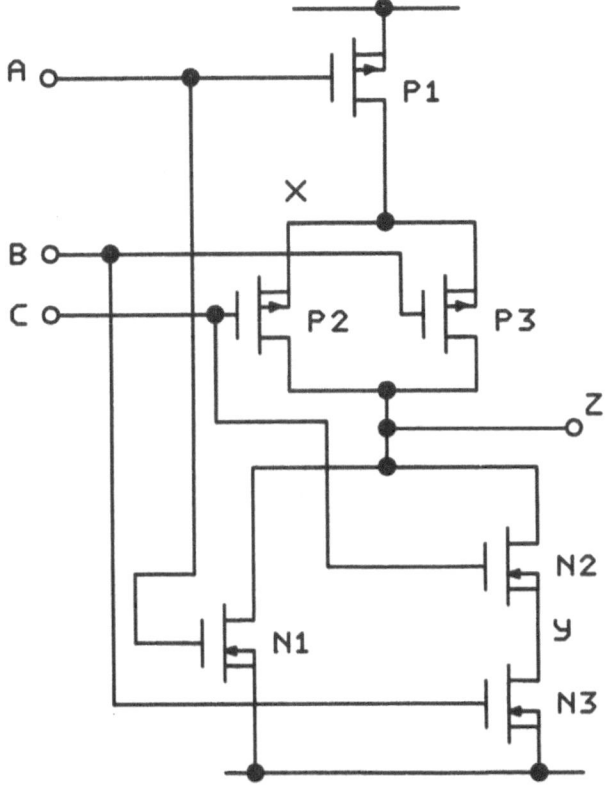

Figure 3.7. 3-Input NOR-NAND gate

Table 3.2. The State Table of the Gate Shown in Figure 3.7

A	B	C	X	Y	Z
0	0	0	1	1/2	1
0	0	1	1	1	1
0	1	0	1	0	1
0	1	1	1	0	0
1	0	0	0	0	0
1	0	1	0	0	0
1	1	0	0	0	0
1	1	1	2/3	0	0

Table 3.3. The State Tables of Transistors P3 and N2
in Fig. 3.7

	P3			N2		
	S	G	D	D	G	S
1	1	0	1	1	0	1/2*
2	1	1	1	1	0	1
	1	0	1*	1	1	1
3	1	1	0	0	1	0
4	0	0	0	0	0	0
5	0	1	0	0	0	0*
	0	0	0*	0	1	0*
6	2/3	1	0*	0	1	0*
	(a)			(b)		

* Redundant states in testing stuck-at type defect

Table 3.4. The Test Set of the Gate Shown in Fig. 3.7 in
Fault-oriented Digital Testing

	A	B	C	Z	Fault coverage
1	0	0	0	1	A/1, Z/0
2	0	0	1	1	B/1, Z/0
3	0	1	0	1	C/1, Z/0
4	1	0	0	0	A/0, Z/0
5	0	1	1	0	B/0, C/0, Z/0

similar to that of Vector 6 of P3 and the worst case here is provoked by Vectors 2, 3 and 4. Vector 4 is included for the benefit of testing shorts between channel and gate.

State 6 in (a) and State 1 in (b) are less crucial compared with others as far as the detection of the defects is concerned since the two transistors are stressed less in these states. It can be shown that such states in all basic gates are not needed in the test set to stimulate the defects. Hence, these states can be deleted from the test set for testing stuck-at type of defects.

Now let us look at the test set needed to detect the single stuck-at faults in the path sensitization methods for the gates shown in Figure 3.7. A complete test set is listed in Table 3.4.

The Vectors 1, 4 and 5 comprise the set for the 2-input NOR gate and the Vectors 2, 3 and 5 comprise the set for the 2-input NAND gate. It can be deduced that all the states of P3 and N2 are covered in the test set of Table 3.4 and, more generally, the states that are needed to detect the single stuck-at type of defects in the StCT for combinatorial logic circuits are covered by the test set of the path sensitization method in digital testing.

However, is it necessary to use the complete test set of digital testing for the StCT? The answer is yes. For simplicity, consider a 3-input NOR gate as is shown in Figure 3.8, with the test set of the path sensitization method listed in Table 3.5. It seems that this complete set is oversized for testing the stuck-at type of defects in the StCT since two vectors might be enough; one sets all the inputs to logic-0 to test for $A/1$, $B/1$, $C/1$ and $Z/0$ and the other sets all the inputs to logic-1 to test for $A/0$, $B/0$, $C/0$ and $Z/1$. Note, the '$A/1$' means the stuck-at type of defects at node A, which does not mean the digital stuck-at. However, this is not an adequate set for the StCT since nodes X and Y, the inaccessible nodes, are not tested at full supply voltages in the first vector. The nodes X and Y are tested at full supply voltages only when the first three vectors of Table 3.4 are used. As a matter of fact, P2 and P3 have to be turned ON to pull node down to the V_{ss} supply rail when input A is tested against $A/0$, and P1 and P3 to ON when input B is tested against $B/0$ so that nodes X and Y are, at the same time, tested against full logic swings.

Therefore, the concept of path sensitization in fault-oriented methods of digital testing is actually consistent with the concept of considering the maximum stress at the inaccessible nodes such as nodes X and Y discussed above in the StCT, that is, the test set of Table 3.5 is necessary for the StCT to test single stuck-at type of defects in the NOR gate shown in Figure 3.8. It can also be proven that the full test sets of any basic gate are exactly the same in both digital testing and the StCT.

The transmission gate is the other type of gate implemented in CMOS technology. Its basic structure is shown in Figure 3.9 and the test set of digital testing for single stuck-at fault is shown in Table 3.6.

Note that the sequence of applying the test vector to the transmission gate has to be obeyed during the test because of the nature of the storage effect.

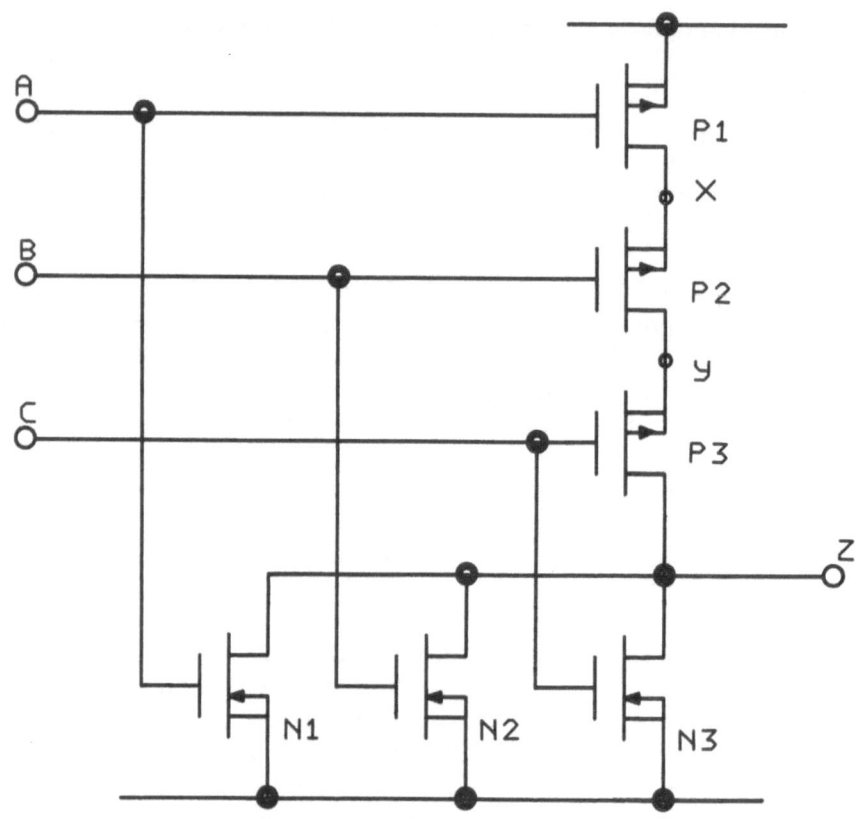

Fig. 3.8.   3-Input NOR Gate.

Table 3.5.   The Test Set of the 3-Input NOR Gate in Fault-oriented
Digital Testing

	A	B	C	Z	Fault coverage
1	1	0	0	0	A/0
2	0	1	0	0	B/0
3	0	0	1	0	C/0
4	0	0	0	1	A/1, B/1, C/1, Z/0

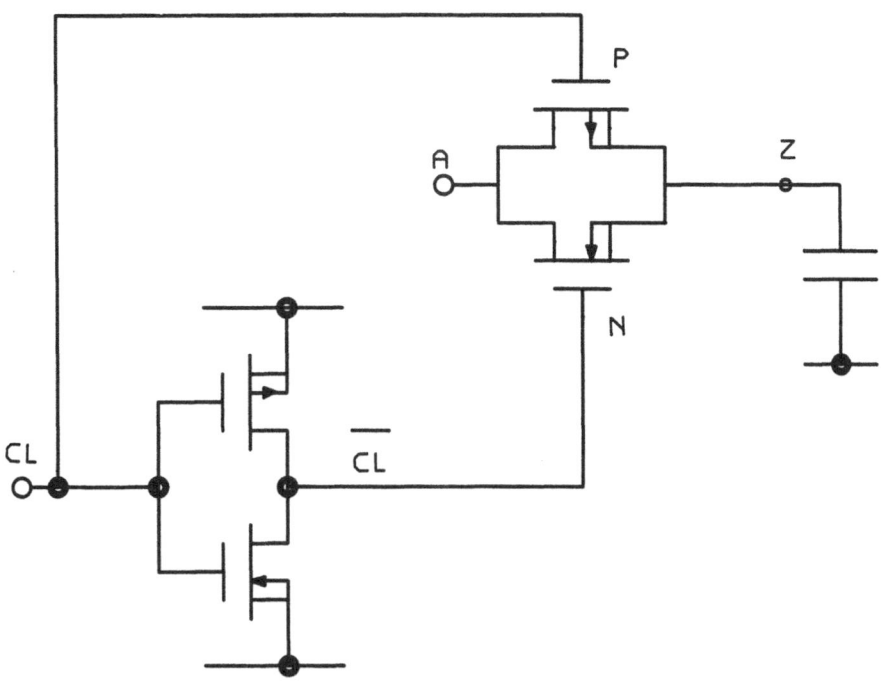

Fig. 3.9. Complementary transistor pair transmission gate together with its driver.

Table 3.6. The Test Set for a Transmission Gate in the Path
Sensitization Method

	A	CL	Z	Fault coverage
1	0	0	0	A/1, Z/1
2	1	0	1	A/0, Z/0, CL/1
3	1	1	1	Preset
4	0	1	1	CL/0

Table 3.7. The Test Set for a Transmission Gate in the StCT

A	CL	Z	Fault coverage
1	0	1	A/0, C1/1, Z/0
0	0	0	A/1, CL/1, Z/1
0	1	0	A/1, CL/0, Z/1

Also, the CL/0 is not detectable in digital testing because the transmission gate has a redundant transistor and therefore a redundant inverter to produce the inverted clock logic so far as performing the logic function is concerned. Because of these characteristics, the test set of the StCT will be different from that of digital testing. As we mentioned before, the 'defect signal' does not have to be transmitted to the output of the circuit in the StCT, so the preset stage can be deleted. The test set may have the form listed in Table 3.7, where the number of vectors is less than that of Table 3.6. In addition, the untestability of the CL/0 in digital testing is overcome in the StCT since the stuck-at-0 type of defect at CL is stimulated by setting CL to 1.

It should be clear that the test set of the path sensitization method in digital testing is adequate when it is applied to the StCT for testing single stuck-at type of defects in both combinational and sequential circuits, although the complete set is not necessary for the StCT when the sequential elements and redundancy circuits are involved. Therefore the techniques developed for test-pattern generation for stuck-at faults are directly applicable to StCT and, even more, the test set of digital testing can be directly used in the StCT without any modification, although the set usually can be greatly reduced in the StCT.

For historic reasons we did not follow the above technique to derive the test sets of the StCT for devices 4013 and 4014 since the set derived from the function-oriented method is, on the one hand, relatively small and, on the other hand, very useful for testing other types of defects such as bridging faults and locating the defects for research interests and failure analysis. The test sets that were derived from the function-oriented method for both 4013 and 4014 shown in Figures 2.2 and 3.17 are listed in Tables 3.8 and 3.9. Note that the set for the 4013 is a complete one that includes all states of the entire circuit, whereas the set for the 4014 is an incomplete one that only includes all the states of each basic gate, respectively, with the internal nodes listed also. The initialization of the 4013 is carried out in the first vector by setting both SET and RESET inputs to logic-1, whereas the initialization of the 4014 is first carried out by vector 01 before the supply current test is started. We may notice from Tables 3.8 and 3.9 that some test vectors are the same if we only look at the external inputs. This indicates the nature of sequential logic circuits. In addition, the set of 4014 is only half the size of the 4013 set, although the 4014 is almost eight times as large as the 4013 in the size of the circuit. This reveals, to some extent, the feature of the StCT, although the test set of the 4014 is rather incomplete. There are probably two factors associated with it: one is that the eight flip-flop stages are tested in parallel and the other is that no transmission of the 'fault signal' to an external output is involved because we only look for the excess current on the power supply.

So far we have discussed the procedures for deriving the test set for detecting the excess supply current which results from the single stuck-at type of defect in the StCT. The so-called stuck-at type of defect is actually the defect that results in abnormal conductance between one of the supply rails and the node that is

Table 3.8. The Test Set of the 4013 D-type Flip-flop in the StCT.
The internal nodes are shown in Figure 3.17.

Vector	Inputs	Internal node states	Outputs
	CL D S R	A B C D E F	Q Q̄
01	1 1 1 1	0 0 0 0 0 0	1 1
02	1 1 1 0	1 0 1 0 1 0	1 0
03	1 1 0 0	1 0 1 0 1 0	1 0
04	1 1 0 1	0 1 0 1 0 1	0 1
05	1 1 0 0	0 1 0 1 0 1	0 1
06	1 0 1 1	0 0 0 0 0 0	1 1
07	1 0 1 0	1 0 1 0 1 0	1 0
08	1 0 0 0	1 0 1 0 1 0	1 0
09	1 0 0 1	0 1 0 1 0 1	0 1
10	1 0 0 0	0 1 0 1 0 1	0 1
11	0 1 1 1	1 0 0 0 0 0	1 1
12	0 1 1 0	1 0 1 0 1 0	1 0
13	0 1 0 0	1 0 1 0 1 0	1 0
14	0 1 0 1	1 0 0 1 0 1	0 1
15	0 1 0 0	1 0 1 1 0 1	0 1
16	0 0 1 1	0 0 0 0 0 0	1 1
17	0 0 1 0	0 0 1 0 1 0	1 0
18	0 0 0 0	0 1 0 0 1 0	1 0
19	0 0 0 1	0 1 0 1 0 1	0 1
20	0 0 0 0	0 1 0 1 0 1	0 1

Table 3.9. The Test Set of 4014 Shift-Register in the StCT.

Vector	Input CL P/S Pin Sin				Internal states A4 A5 A6 A7 A9					Output Q6 Q7 Q8
00	0	1	1	0	0	1	1	0	0	1
01	1	1	1	1	1	1	1	0	0	1
02	1	0	1	1	1	1	1	0	0	1
03	0	0	0	1	1	1	1	0	0	1
04	0	1	0	1	1	0	0	1	0	1
05	1	1	0	0	0	0	0	1	1	0
06	1	0	0	0	0	0	0	1	1	0
07	0	0	1	0	0	0	0	1	1	0
08	0	1	1	0	0	1	1	0	1	0
09	1	1	0	1	1	0	1	0	0	1
10	0	1	0	1	1	0	0	1	*	*
11	1	1	0	0	0	0	0	1	1	0
12	1	1	1	0	0	1	0	1	1	0

* Undefined states.

set to the opposite supply rail. Such abnormal conductance can exist between any two physically adjacent nodes. In a less severe case, the defects as such produce excess supply current and, in a more severe case, cause malfunction of the circuit. By its nature, one calls such a fault a bridging fault. The stuck-at fault model in digital testing is not adequate for covering bridging faults, and to include the bridging faults in the fault list usually complicates enormously the derivation of the test set. We present a short discussion about bridging faults below.

A bridging fault in digital testing is a short between the two nodes in a circuit, and a bridging defect in the StCT is defined as the abnormal conductance between the two nodes in a circuit. The difficulty arises when we try to test this theory because any two nodes in a circuit can, in principle, be bridged. In mathematical terms, a complete fault list for a circuit with n nodes would have $2n(n-1)/2$ such faults. To test for a fault coverage of 100% we have to include this number of faults in the list if the detailed layout of the circuit is unknown.

It is, first of all, impractical to apply such algorithms to VLSICs since deriving a test set for such a complete fault list would involve millions of pairs of nodes for the circuit. Fortunately it is not necessary to consider all pairs of nodes in order to derive the test set since an accidental short between tracks is unlikely to occur unless the two tracks are physically adjacent. Therefore, the derivation of the test set for bridging defects between two nodes could be practical if a detailed layout of the circuit is known, especially if the testability of such a fault is considered in the design of the circuit, or only the bridging faults at the inputs and outputs are considered if the detailed layout of the circuit is unknown.

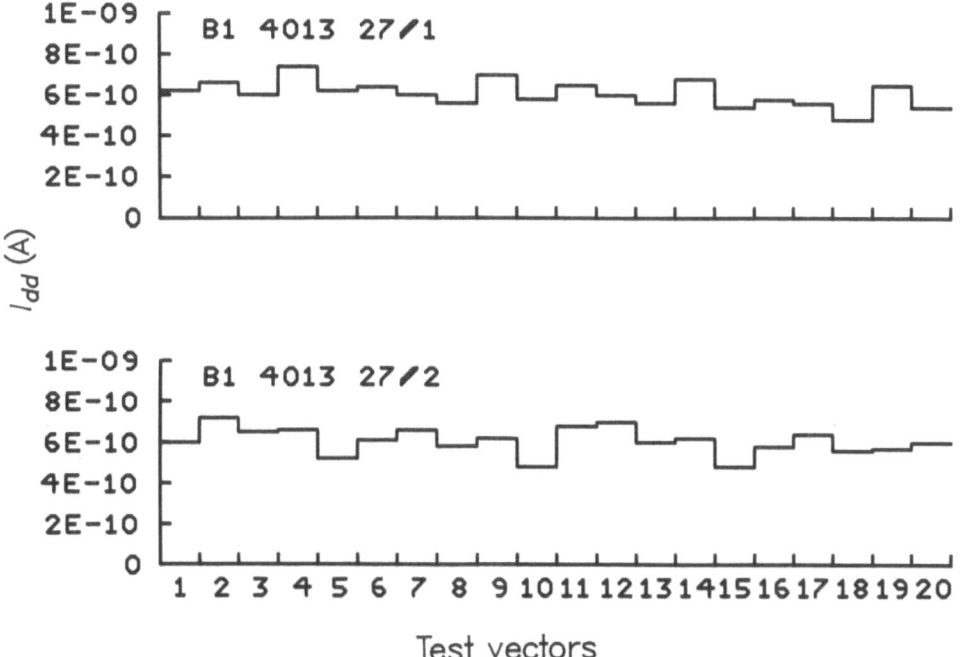

Fig. 3.10. The test static supply current of a normal D-type flip-flop versus test vector for the two devices of the dual 4013 number 27.

However, we have taken bridging defects into account when deriving the test set for 4013 and 4014 since all the states of the circuit are tested against the excess supply current, ie, all bridging defects that result in excess supply current are tested for in the test set. Note that bridging defects between the two flip-flops are not accounted for because such defects are very rarely found in real life.

3.3.1.3. Features of the StCT

As was stated previously, the intrinsic supply current of the CMOS circuit is contributed mainly by the reverse biased p–n junctions between the substrate and the well-substrate of the transistor when the circuit is in static operation and therefore, the amplitude of this current does not vary much with the input and internal state vectors. General characteristics of the current against input vectors are illustrated in Figure 3.10. The two parts of the plot illustrate the dependence of the total supply current on the input vectors of each of the two devices on the same chip. The current is quite consistent over all vectors and even in both plots. This manifests our assumption in deducting the intrinsic supply current. A further look at the $I \sim V$ curve will give an even better explanation of this. The $I \sim V$ curve of B1 4013 05/1+2 at vector 20 is shown in Figure 3.11. The dotted curve is obtained from the measured data. As predicted by the generation current theory for the silicon p–n junction, the static supply current is about 4.7×10^{-10} A at 14.8 V in this sample, and varies with a power of the supply voltage. The reverse generation current law $I = a(V_{bi} + V)^{1/2}$ is fitted to the measured data between 3 and 14.8 V and this is shown by the solid curve in the diagram. The exponent in this example is 0.58 which is a little bit larger than the predicted value 0.5 for an abrupt junction with the coefficient of the determination, r^2 very near to '1', which indicates a high quality of fitting achieved by this regression.

It has been observed that the static supply current of the devices in the same manufacturing batch usually show a good consistency in both magnitude and shape of the $I \sim V$ curve even though a relatively large deviation, say about two decades, has been observed in the static supply current between similar devices made by different manufacturers. However an exception to this has also been observed for devices in Batch 4, which we will discuss at a later time.

3.3.2. The Cut-off Frequency Test

3.3.2.1. Introduction to the CoFT

The cut-off frequency is, in the context of this book, defined as the frequency above which (and for dynamic memory devices perhaps also 'below which') the digital circuit fails to perform the expected functions at a given supply voltage. The physical description of this definition is as follows. When a digital circuit (suppose a static device) such as a shift-register works at a given voltage and the input clock frequency of the device is gradually increased, some particular cell in the circuit will stop responding effectively or not at all at some input clock frequency. Therefore the circuit output stage will not show the expected signal when the fault is propagated to it. We then call this frequency the 'cut-off frequency'. By 'fault' in this context we do not necessarily mean a digital fault of the circuit. In fact, the interpretation of the 'fault' in this context depends on

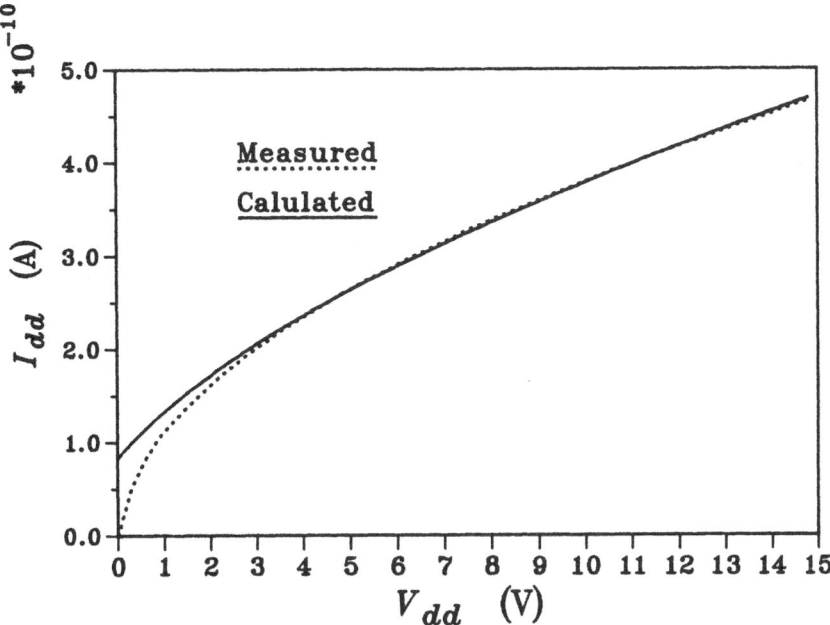

Fig. 3.11. The characteristics of the static supply current versus supply voltage for a normal D-type flip-flop circuit.

how the circuit behaves and where the circuit is used. There are two other situations where the 'fault' may be considered in addition to the most obvious definition, the functional fault appearing at the tested output of the circuit. Excess delays of the output signal also suit our definition in some cases. For example, in a circuit board where a chain of D-type flip-flops are connected with a common clock signal to perform the function of a register, the delays of each flip-flop should be within the period of the clock signal to permit correct transfer of the data from stage to stage. Therefore the cut-off frequency of the flip-flop in this case may be defined according to the delay of the signal from input to output, provided the functional operation is still correct. Besides the functional failure and the excess delay, we may encounter another situation where the circuit produces a degraded performance, for example, a decreased logic swing of the output waveform. The cut-off frequency in this case is defined as the frequency when the logic swing of the output waveform falls to a predefined value. To illustrate, we simulate the circuit shown in Figure 3.12 by using PSPICE. The device dimensions are also shown in the diagram, which were taken from the 4013 of 'National Semiconductor'. The simulation parameters are estimated based on the results of simulation work in Chapter 2. The response of the circuit to a 10 MHz input signal is shown in Figure 3.13, where the waveforms of three outputs are displayed together with the input signal. The dashed curve is from the output of a chain of ten identical inverters and is clearly out of phase with

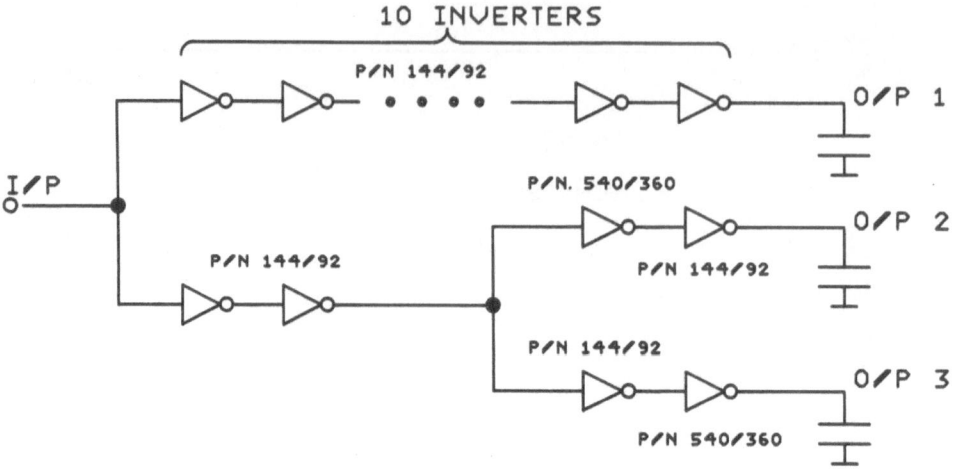

Fig. 3.12. The schematic circuits used to simulate the cut-off characteristics.

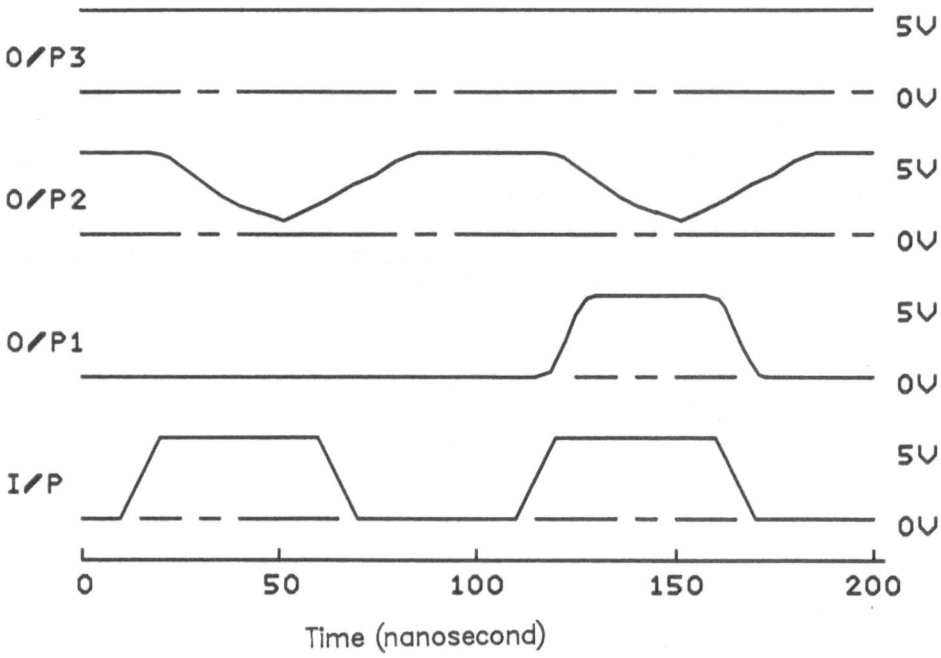

Time (nanosecond)

Fig. 3.13. The resultant waveforms obtained from the simulation on the circuit shown in Fig. 3.12 together with the input waveform.

the input signal shown by the solid curve, although it still performs the expected function. If this happens in a sequential circuit, the signal at the output will be misinterpreted when both input and output are latched by the same clock signal and therefore will cause a malfunction of the circuit. The dotted and the chaindotted curves are from the outputs of two otherwise similar circuits, except for the two inverters with different dimensions being connected in different order in the two paths. Because of the matching, Path 2 does not produce any signal at the output, whereas Path 3 does. However, Path 3 still does not produce a full logic swing at the output. Care is therefore needed when we refer to the cut-off frequency of an operating digital circuit.

As mentioned earlier, CMOS technology has negligible static current flowing from one logic gate to the other. The logic function is performed by charging and discharging node capacitances by the available drain current of an ON transistor. The basic features of the charging and discharging processes are illustrated by the NOT gate shown in Figure 3.14(A) and (B) respectively. We will show in the next section that the charging to the node capacitance C_{node} is performed only by the p-channel transistor and discharging by the n-channel transistor. The 'node capacitance' C_{node} is the equivalent capacitance of all the capacitances asscociated with that node, as shown in Figure 3.15. Thus the cut-off frequency as defined reveals the process of active charging and discharging in each of the elements that comprise the circuit path being tested and therefore the availability of the drain current in both p-channel and n-channel transistors to perform such a process.

The implementation of the CoFT may be done by the system shown schematically in Figure 3.16. The pattern generator is used to exercise the circuit and the output signal is analyzed against one of the criteria discussed above to determine the cut-off frequency. In addition, the high-speed analog buffer is used to produce a better performance for the test since unpredicted loading of the device being tested affects its output stage. Also, a precision power supply with variable voltages is used to provide several test voltages. This will be further discussed later in the book.

To derive such a test we need to generate the exercise pattern first as we did for the StCT. In the following subsection a brief discussion is given on how we decide on the specific circuit paths to be tested and how we can generate patterns that efficiently test the circuit.

3.3.2.2. Test Pattern Generation for the CoFT

The time needed for the process of charging and discharging discussed above is the delay time of a unit that performs the process. The summation of all the delay times associated with the units that comprise the circuit path being tested determines the cut-off frequency of the circuit portion in the path. Thus the basic principle of selecting the test path is to include as many units as possible in the path so that more units are evaluated in one operation. A long delay

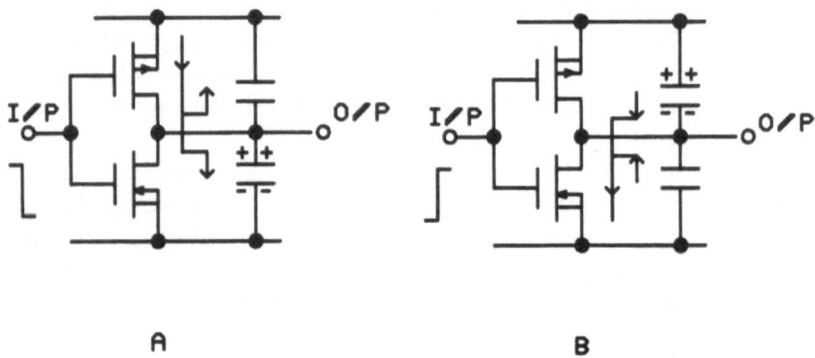

A

B

Fig. 3.14.   The charge and discharge process of a basic inverter when it is changing its logic states.

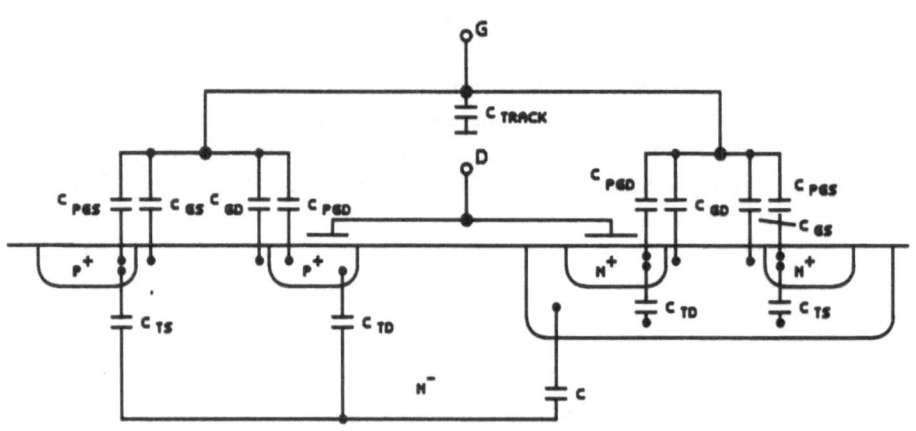

Fig. 3.15.   The capacitances of a basic CMOS inverter.

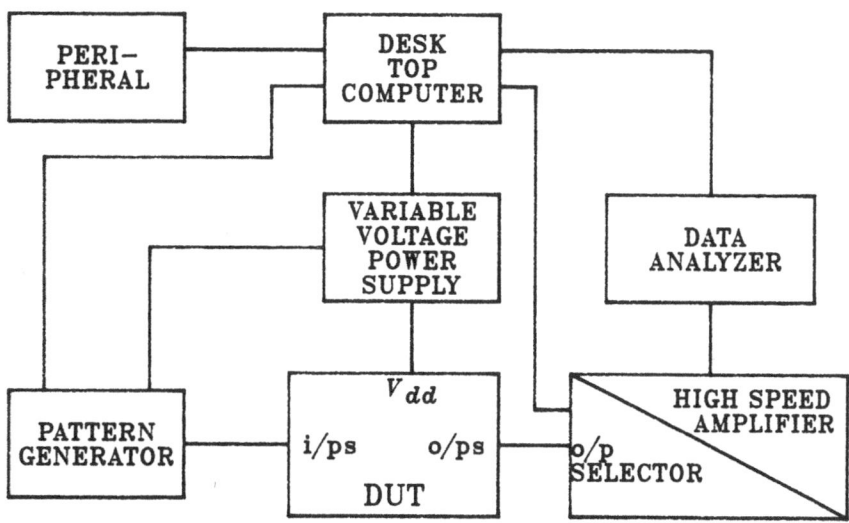

Fig. 3.16.   The block diagram of the CoFT system.

indicates a lower cut-off frequency and therefore the demand on high-speed test equipment should be less severe. A long path, however, makes the discrimination of a single anomolous delay in the path more difficult to detect.

In addition to the above principle, we also have to consider how the delay time is developed so that some critical elements can be tested if we aim to test for the lowest operating frequency.

After the selection of the test path we then need to sensitize such a path to permit the signal to be propagated to the output. This can be performed by using the concept of the path sensitization mentioned in the previous subsection. All the basic units along the path are first sensitized and all other parts of the circuit are then set to meet the requirement of the sensitization of those units along the path.

To illustrate the procedure of the pattern generation for CoFT we consider the 4013 and 4014 as examples. The 4013 is a sequential circuit with the features of combinational logic. The circuit diagram and transistor dimensions are shown in Figure 3.17 for a 'National Semiconductor' device. For simplicity we consider first of all the circuit as combinatorial logic with the transmission gate left short or open depending on their states. There are many different paths from the input SET or RESET to outputs $\bar{Q}$ and Q. Short paths are produced between the SET and output Q through the first or fourth NOR gate depending on the state of clock input, and between the RESET output $\bar{Q}$ through the third NOR gate. Long paths are produced between SET and $\bar{Q}$, and the RESET and Q through two NOR gates also depending on the state of the clock input. The simulation

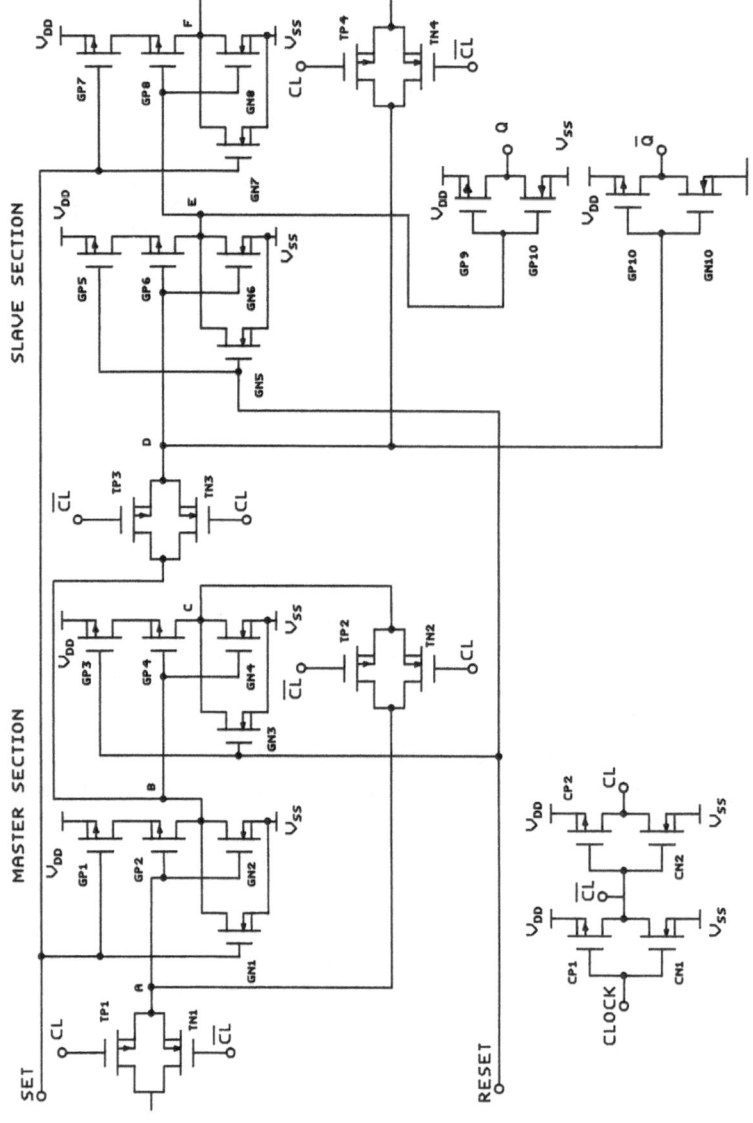

Fig. 3.17. The transistor-level circuit diagram for the D-type flip-flop made by a specific manufacturer with the measured device size labeled.

has shown that a great difference exists in cut-off frequency between these short and long paths which is, of course, expected. To test for a weak device, long paths will be more suitable since more transistors are tested and low operating speed is expected and therefore the demand on the high speed equipment is less. To characterize the individual transistors for the interest of research, short paths are better since fewer elements are involved.

To test either a short path or a long one we need to sensitize the path so that the test signal will be permitted to transmit along the path and build up at the output where the signal is detected, as we mentioned earlier on. In the case of the above circuit, the only multi-input gate is the NOR gate. To test the 2-input NOR gate, one input terminal needs to be set to logic-0 while the other input terminal is tested. For example, the node A and input RESET have to be set to logic-0 while the path between SET and $\bar{Q}$ through the first and third NOR gates is tested.

The test suggested above only tests half of the NOR gate. To test the other pair of transistors in the gate, sequential logic has to be performed. That is, its basic toggle operating mode is performed. In this mode we may disable all the asynchronous inputs to sensitize the toggle path and simply clock the data from the D input to the outputs to test the path. The basic structure of the test pattern can have the form as shown in Figure 3.18. The pattern of input D is a square wave with a frequency equal to half of the clock signal. Note that the phase relationship between these two patterns is very important since this specific type of device is positive-edge triggered.

The switching efficiency of the transmission gates is crucial in the toggle operating mode of the devices. The clock signal in a sequential circuit is usually working at a higher speed compared with other signals and it is not uncommon to have a clock signal to drive a number of gates, especially transmission gates. Thus, a circuit like a D-type flip-flop can very possibly suffer from a deficiency in charging process through the transmission gates, as has been discussed previously in Chapter 2.

Again we have to decide which output is to be monitored. The output $\bar{Q}$ gives a slower operating response than does the output Q in this operating mode since an extra NOR gate is involved in output $\bar{Q}$. In a simple test the output $\bar{Q}$ is a better choice. To achieve better performance for the test, both outputs should be tested and compared.

As has been indicated previously, the 4014 shift register is basically a chain of 4013 D-type flip-flops. The main difference between these two devices is that the 4014 does not have the asynchronous set and clear facilities. Thus only sequential logic is tested and therefore only the toggle mode is applicable in the 4014. The 4014 has a common clock signal amongst its eight stages. The simulation in Chapter 2 suggests that the enormous delay of the clock may be the main obstacle to the high-speed operation of this device. The moderate degradation of the signal both in amplitude and delay in a specific stage is usually corrected by the succeeding stage and therefore only the clock circuitry and last stage are tested in the toggle mode for normal devices. As far as the testability in the AC operating mode is concerned, this device is not well designed. Historically, the test pattern of the CoFT for the 4013 was derived as shown in Figure 3.19(A), (B) and (C). The toggle operation is tested by the pattern shown in diagram A, where the asynchronous inputs are disabled. The asynchronous inputs SET and RESET are tested together with the toggle operation by the patterns shown in diagrams B and C, respectively. It is important to understand the effect of the inputs SET and RESET on the toggle operation since they asynchronously set and reset the device independent of the toggle operation and therefore the position where they are inserted needs to be considered. Further discussion of this will be given in the next subsection, together with a consideration of their cut-off characteristics.

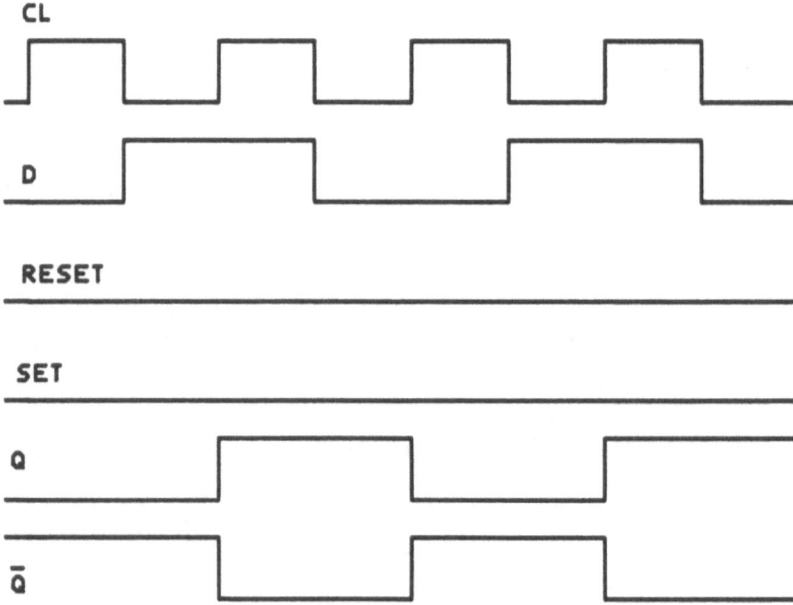

Fig. 3.18.   The test pattern of the CoFT for D-type flip-flop in the toggle mode.

The derivation of the test pattern for the 4014 shift-register is similar to that of the 4013. The 4014 does not have the asynchronous set or clear facilities so that combinational logic does not exist. Thus we only test for toggle operation of clock and serial data input. The pattern is as shown in Figure 3.19(A) for 4013.

68

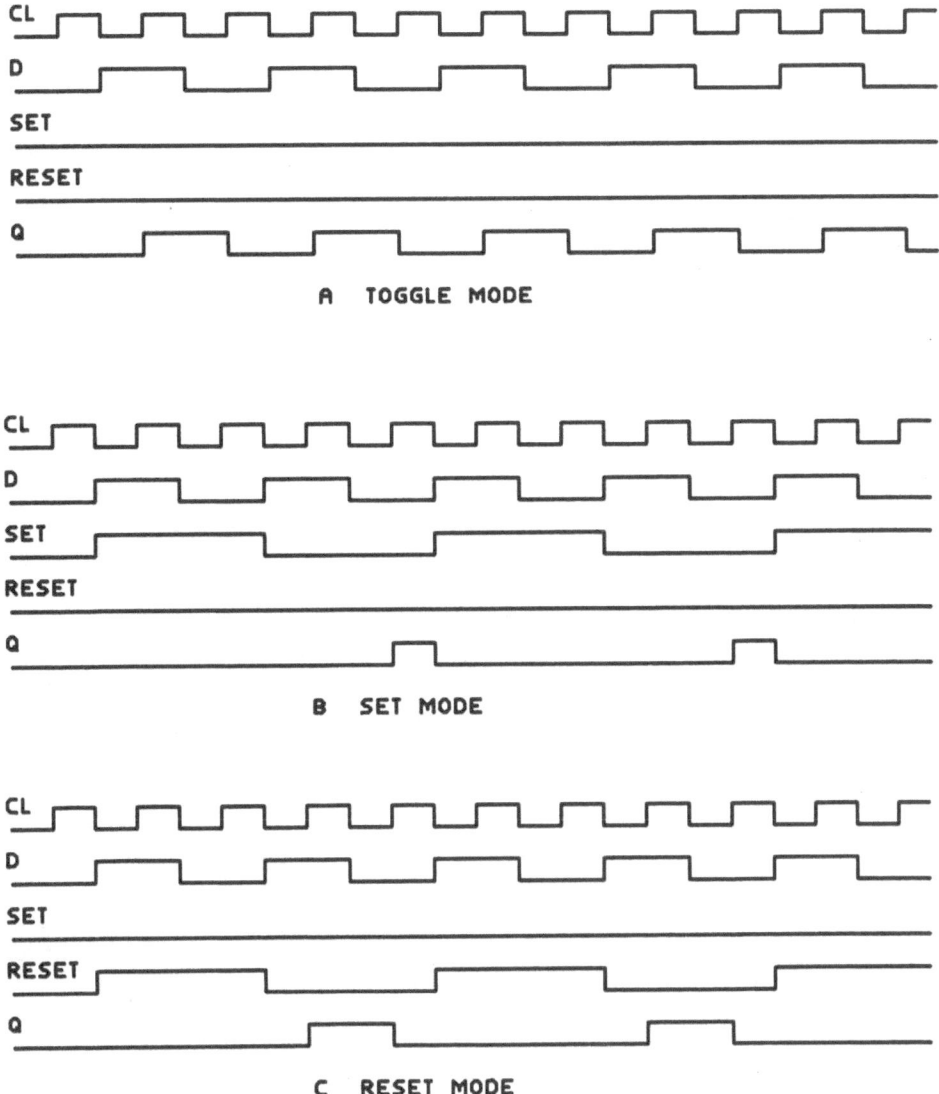

Fig. 3.19.   The test pattern of three operating modes in C₀FT for the D-type flip-flop.

3.3.2.3. Features of the CoFT

It has been pointed out that the limits of the cut-off frequency of the CMOS circuit originate from the time needed to charge the node capacitances along the path being tested by the available drain current of the transistors. The charge process of the CMOS circuit has, in general, the form of Equations 3.3 and 3.4

$$t_{fall} = \frac{2\,C_{out}\,L}{Z\,C_{ox}\,\mu_n\,V_{dd}} \left\{ \frac{v_n - 0.1}{(1 - v_n)^2} + \frac{tanh^{-1}\left(1 - \frac{0.1}{1 - v_n}\right)}{1 - v_n} \right\} \qquad (3.3)$$

$$t_{rise} = \frac{2\,C_{out}\,L}{Z\,C_{ox}\,\mu_p\,V_{dd}} \left\{ \frac{v_p - 0.1}{(1 - v_p)^2} + \frac{tanh^{-1}\left(1 - \frac{0.1}{1 - v_p}\right)}{1 - v_p} \right\}, \qquad (3.4)$$

where $v_n = V_{Thn}/V_{dd}$ and $v_p = V_{Thp}/V_{dd}$. Both the fall and the rise times, t_{fall} and t_{rise}, are inversely proportional to the power supply voltage V_{dd} and therefore the cut-off frequency is proportional to V_{dd}. Physically, this is understandable since the available drain current is determined by the channel conductance or transconductance, which in turn are dependent on the gate and drain voltages. The general experimental feature of the cut-off frequency versus supply voltage is plotted in Figure 3.20 for a normal 4013 D-type flip-flop. The results obtained from the three test patterns discussed in the previous subsection are displayed in the diagram. It can be seen from the diagram that a very linear dependence exists in each case, although the relative amplitudes are quite different for the different test patterns.

Thus it is not necessary to conduct the test at all voltages since a linear dependence exists between V_{dd} and the cut-off frequency. However, what voltage should be chosen for the test? This may be justified with respect to two aspects. One is that the drain current will be affected more by the threshold voltage of the transistor when the supply voltage is low. Secondly, the circuit will cut-off at a lower frequency when a low supply voltage is applied. This is especially desirable when high-speed devices are tested. An optional test at high voltages is also useful since some failure mechanisms are only provoked at high voltages, for example, junction breakdown. Tests done at low voltage may not account for such failure mechanisms, although they are believed to be related.

The large discrepancies between the cut-off frequency appearing in the different operating modes exist because of the different paths and timing associated with them. In the set mode, the input SET is inserted together with the toggle operation. As shown in Figure 3.19B, the LOW to HIGH transition of output Q by toggling only lasts half of the clock period due to the succeeding insertion of the input SET. If the circuit works at a slow operating speed, the discharging process of output Q can be completed before the insertion of the SET. However, while the circuit is working near the maximum operating frequency, the above process cannot be completed before the insertion of the SET in half of the clock cycle. In addition, because of the relative long delay of the toggle operation

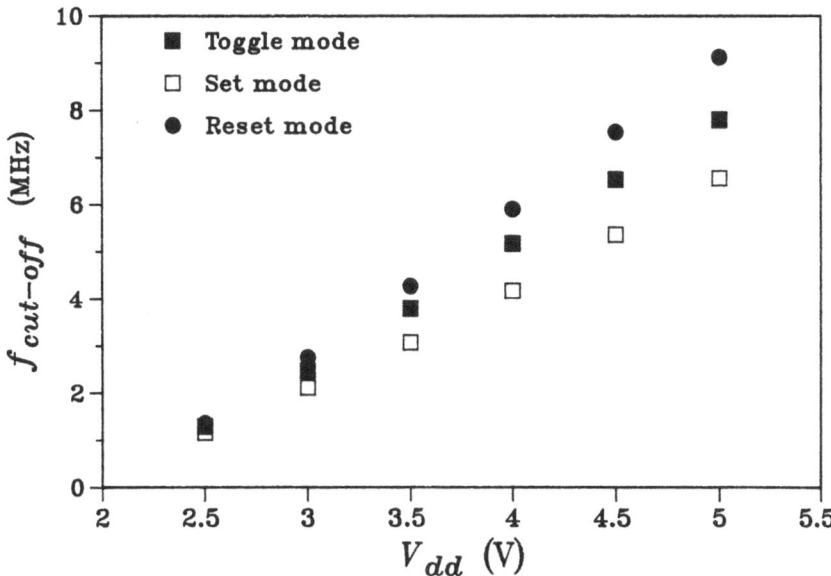

Fig. 3.20. The characteristics of the cut-off frequency versus supply voltage for normal D-type flip-flop.

and short delay of the path from SET to Q, the above process will even be incomplete. This inevitably causes the associated path working under the constraints of premature charging/discharging process, and therefore the cut-off frequency of such path, to decrease. The simulation has also shown that the cut-off frequency is 5–10% higher in the toggle mode than in the set mode. This is consistent with the experimental results, although the experimental data did not show as large a difference as expected both by the above analysis and simulation. In the reset mode, the input reset is inserted to the toggle operation. As is shown in Figure 3.19, the output waveform is otherwise the same as in the toggle mode except for the part where Q is forced to logic-0 when the RESET is inserted. That is, a complete process of the charging and hold at Q is performed, which is different from that of the set mode. Thus, a similar value of the cut-off frequency as in the toggle mode should be obtained since the operation path from RESET to Q is usually faster than the toggle operation. The simulation has also shown a similar result. However, as is shown in Figure 3.20, the cut-off frequency in the reset mode is far higher than that of other two operating modes and this has been observed in all 4013 devices which we have studied. The cause of such a large discrepancy is unknown. It may be caused by the asymmetry in quality of transmitting logic-0 or logic-1 along the path being tested in the toggle mode. If the transition of logic-0 to logic-1 is 'easier' than the opposite transition at Q in the toggle mode, then the cut-off frequency in the reset mode can be higher than that of the toggle mode since the time permitted for the '1' to '0' transition at output Q is long enough. In addition, the positive feedback of this device may also play a part in it.

In general, the devices made by a specific manufacturer show quite uniform values of the cut-off frequency under specific operating conditions. Both Batches 1 and 2 have shown less than 20% variation in cut-off frequency. This is not surprising because the main factors that affect the cut-off frequency are the threshold voltages and the gain factors, which do not vary much over the devices made in the same process. This is especially true when the two devices in a chip are compared. A deviation of less than 1% in cut-off frequency between two such devices is normally observed. However, a large deviation in cut-off frequency of devices in Batch 4 has been observed. These are suspected to be not well made devices and will be discussed later.

However, relative deviations of the cut-off frequency usually exist among the devices made by different manufacturers. For example, the cut-off frequency of the specimens in Batch 2 is about 20–50% higher than that of the specimens in Batch 1. This feature may reflect the geometrical properties of the circuit and the characteristics of the MOS transistors.

3.3.3. The Transient Current Test

3.3.3.1. Introduction to the TrCT

When a circuit is exercised to change the state of the set of internal gates, a consequent current pulse is produced which corresponds to the charge flowing through the current paths to change the state. We call such a current pulse the transient current. To illustrate we discuss the charging and discharging process of an inverter gate. Figure 3.21 shows the two parts of the process, where C1 is the equivalent capacitance between the output of the inverter and the positive supply rail originated from the load, and C2 is the equivalent capacitance between the output of the inverter and the negative supply rail. The charging process of the inverter is defined as the input going from high to low, which is shown in Figure 3.21A.

The drain current of the p-channel transistor is the charge process which consists of three components: the current discharging C1, the current charging C2 and the current flowing between the two power supply rails through the two transistors when the transition occurs. The simulation by PSPICE has shown that the third component is negligible so long as the rise and fall times at the inputs are shorter than hundreds of nanoseconds which is usually the case. Thus the charging process is only associated with the discharging of C1 and the charging of C2 and we have

$$I_p = I_2 + I_3.$$

Similarly, the discharging process of the inverter is defined as the input going from low to high, which is shown in Figure 3.21B. Here, the drain current of the

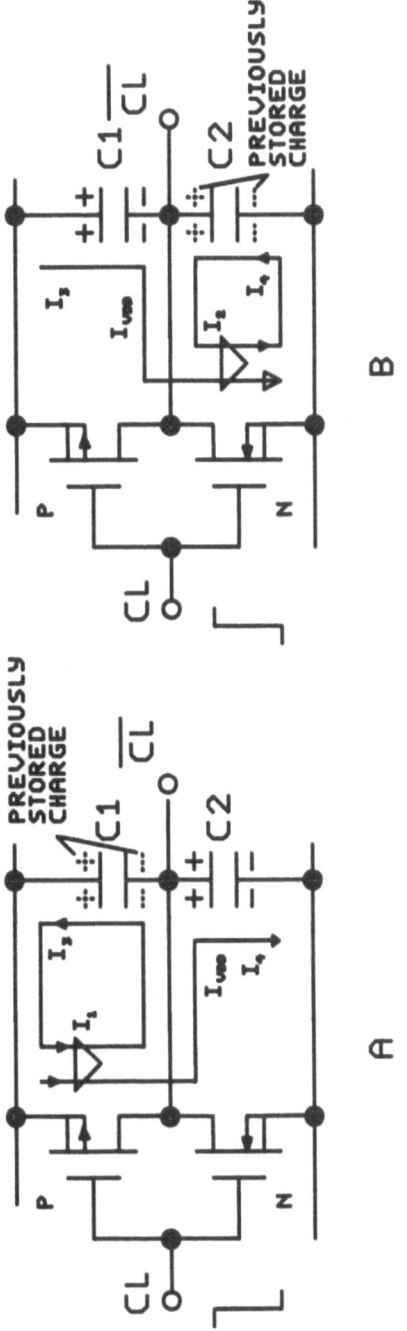

Fig. 3.21. The detailed charge and discharge processes involved in the logic transition of a basic inverter.

n-channel transistor consists of the current of charging C1 and discharging C2. Again we have

$$I_n = I_2 + I_3.$$

It is clear that there is no intersection between I_p, I_n and I_s, the charging process is performed purely through the p-channel transistor and the discharging process purely through the n-channel transistor as regards the inverter output.

Now let us consider the total supply current associated with the charging and discharging process. In the charging process the discharge of C1 is conducted through the closed loop including the p-channel transistor and therefore does not contribute to the total supply current. Similarly, the discharge of C2 is conducted through the loop including the n-channel transistor and does not contribute to the supply current either. So in fact only the charging current of C2 contributes to the supply current in the charging process of the inverter and the charging current of C1 to the supply current in the discharging process of the inverter. We may have noticed that the drain current of the transistor is twice as much as the supply current if we suppose the two transistors are similar and C1 and C2 are equal. This is not surprising since the charge stored on C2 or C1 in a process are discharged in the next opposite process.

Each transition associated with either a charging or a discharging process in an inverter results in a particular transient current waveform and the current pulse is well characterized by the drain current of the MOS transistor in both the linear and the saturation regions. All CMOS circuits are exactly the same as far as the switching process is concerned since the digital transition is always performed by charging and discharging of node capacitances. Thus the transient current pulse of each individual transistor in any circuit results in a particular transient current waveform, even though the individual transient current waveforms may possibly be superimposed on each other.

The transient current test (TrCT) monitors the transient current from the power supply under a certain operation mode and at a given supply voltage. The test system is shown schematically in Figure 3.22. The implementation of this system is similar to that of the CoFT. Basically, the pattern generator generates a test pattern to exercise specific parts of the circuit and the digital oscilloscope is used to record the supply current waveforms. The current to voltage conversion is performed by a 50 Ω resistor. The use of the passive component here does not deteriorate the performance of the DUT and the system since the maximum current produced in the transition is usually in a order of milliamperes and thus the maximum voltage drop across the resistor is less than 0 1 V. The value of 50 Ω is chosen to match the loading of the 50 Ω coaxial cable, and the input impedance of the preamplifier. Unlike the CoFT, a crystal-controlled oscillator is used to produce the clock signal for the pattern generator in the TrCT. This guarantees the stability of the operating frequency and therefore the uniformity of the current pulse waveforms. In addition, an appropriate capacitative loading at the outputs of the DUT is needed to permit the output stage to be tested

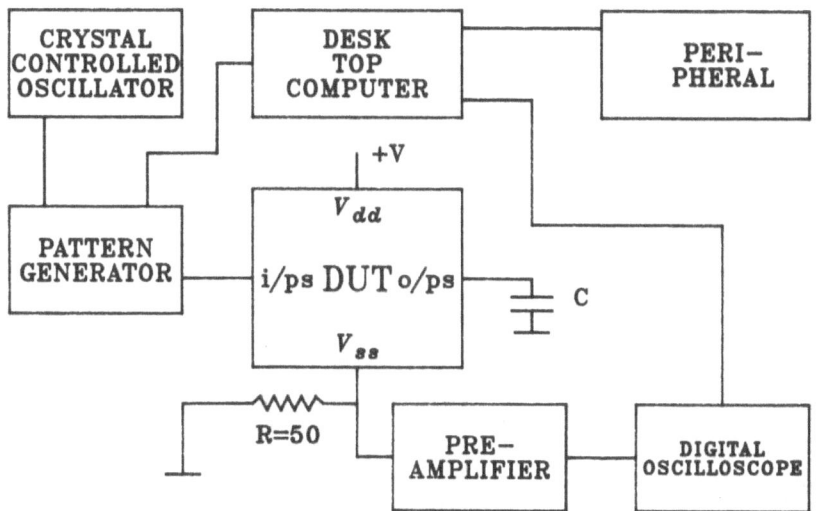

Fig. 3.22.   The block diagram of the TrCT system.

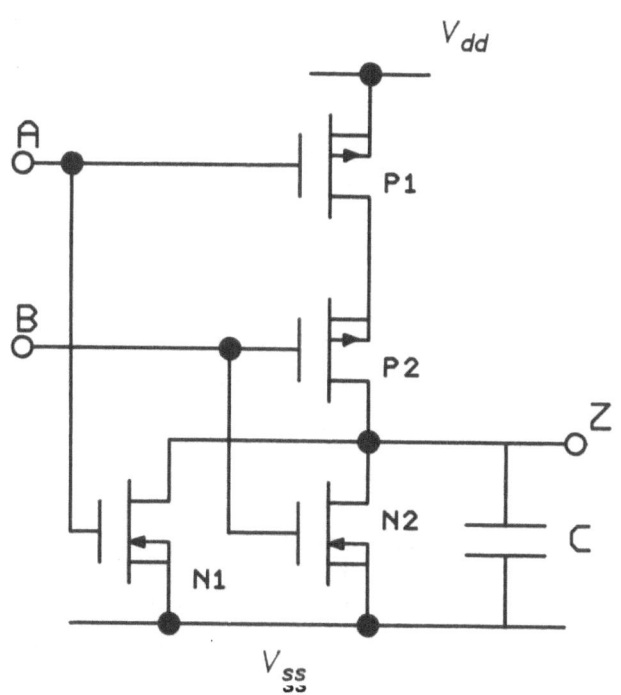

Fig. 3.23.   The 2-input NOR gate with load capacitance C.

realistically provided it does not cause an overwhelming contribution to the total current. Usually, a capacitor with a capacitance of a few picofarads can be used.

In the simulation described in Chapter 2 we have seen that the transient current pulse of each transition in the circuit is an extremely short pulse. Its total width is a few tens of nanoseconds and the rise and fall times are only a few nanoseconds. It is important to use a high-speed digitizing oscilloscope for this test. The one we used is DATA 6000, which has 16 bits amplitude resolution and 200 picosecond time resolution.

3.3.3.2. Test Pattern Generation for the TrCT

The derivation of the test pattern in the TrCT is very similar to that of the CoFT. Basically, we need to exercise each transistor in the circuit by turning it on to charge or discharge the associated capacitance. To achieve this we again need to use the concept of the path sensitization which has been discussed for both the StCT and CoFT. To illustrate, we discuss the procedure of testing the 2-input NOR gate of Figure 3.23. To test one n-channel transistor we need to set the other n-channel transistor OFF so that the discharge of the node capacitance at Z will only be performed by the transistor. Similarly, the one p-channel transistor is kept ON, instead of OFF in the case of n-channel transistors, while the other p-channel tranistor is tested. Thus the procedure for this gate can be as listed in Table 3.10.

To summarize, we plot in Figure 3.24 the input test waveforms and the resultant transient current waveform which is obtained from the simulation by PSPICE on the gate discussed above, where the parameters are extracted from the 4013 D-type flip-flop device made by 'National Semiconductor'. Each transient current pulse corresponds to the ON/OFF transition of a specific transistor. To implement this last sequence we need to generate two input waveforms to exercise the circuit and test for four transient current pulses. We may consider this gate as an inverter, ie, two inputs being connected together, and test it by using a square wave. In this case we need only generate a single exercise waveform and test two transient current pulses. Thus a simpler test may be derived as compared with the former one. However we lose the resolution in this case since two transistors are switching simultaneously. Therefore, it is again obvious that a compromise or optimization has to be made between the efficiency of the test and the resolution.

Historically, the test patterns are derived based on the ones used for the CoFT. For the 4013 D-type flip-flops there are two parts of the test pattern: one which we call the 'toggle mode' and the other which we call 'set/reset' mode. These are depicted in Figure 3.25A and B respectively. For the 4014 shift- register the test patterns are derived based on the individual stages.

The test pattern only consists of the toggle operation in this device since there is no asynchronous input facilities. Each stage is toggled by exercising the parallel in and clock inputs in a pattern shown in Figure 3.26, with each stage isolated by setting input/parallel/serial to parallel operation.

Because of the large memory storage associated with this test, we did not test each stage individually. Instead, we tested the first stage alone, the first five stages together, and the last three stages individually.

Table 3.10. The Test Set Sequence of the TrCT for a 2-Input NOR Gate

	A	B	Z	Supposed initial states of the transistor
1	0	0	1	P2
2	1	0	0	N1
3	0	0	1	P1
4	0	1	0	N2

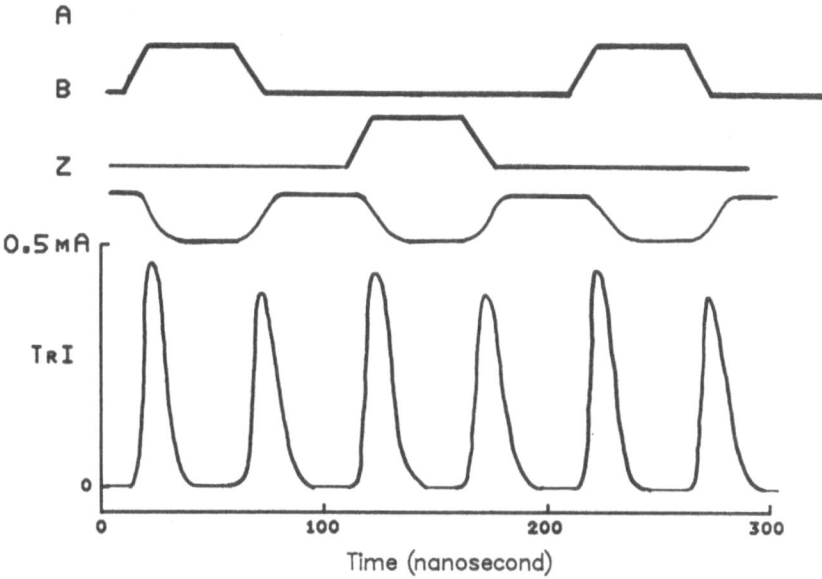

Fig. 3.24. The transient current waveform obtained in the simulation of the circuit shown in Fig. 3.23, with the logic waveforms of the input and output also shown.

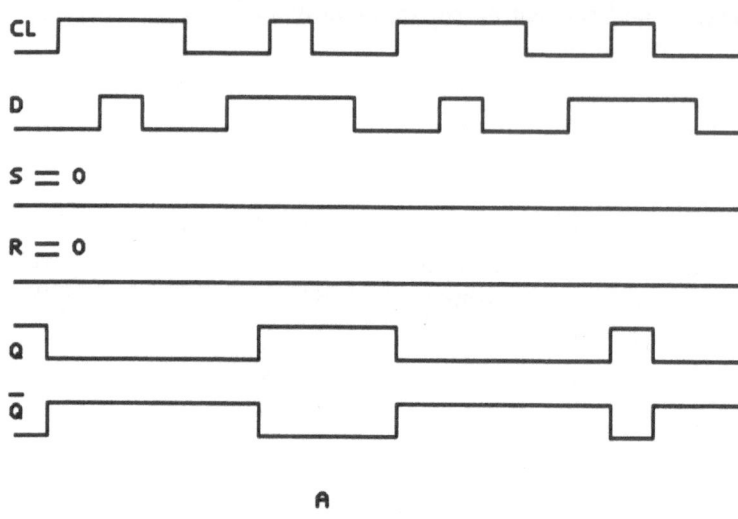

A

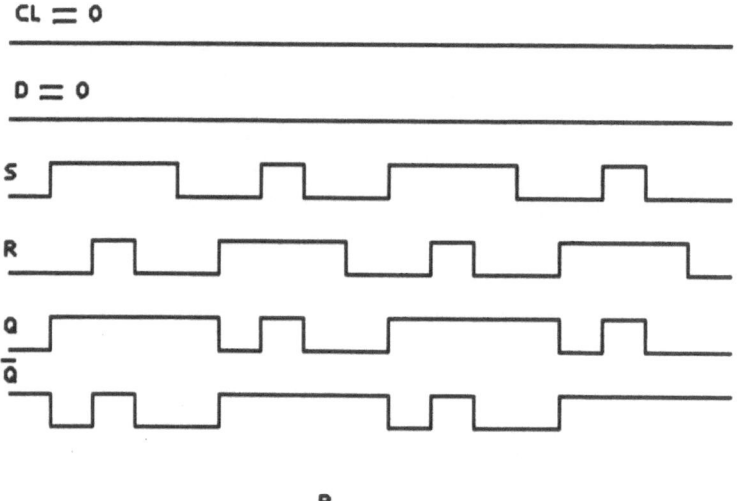

B

Fig. 3.25. The test pattern used in the TrCT for the D-type flip-flop.

3.3.3.3. Features of the TrCT

In the previous subsection we have pointed out that the transient supply current is generated by the process of charging node capacitance through the drain current of the transistor. Each logic transition of the capacitive node generates a specific transient current pulse. By solving the equation $C\,dV/dT = I$ using dt the current I_{dd}, we should be able to obtain such a pulse waveform. In a practical circuit such individual pulses will be superimposed on each other. In Chapter 2 the transient waveforms have been simulated for the 4013 made by 'Motorola'. The experimental results showed very similar results to those predicted by simulation. The transient current waveforms of the 4013 in both toggle and set/reset modes

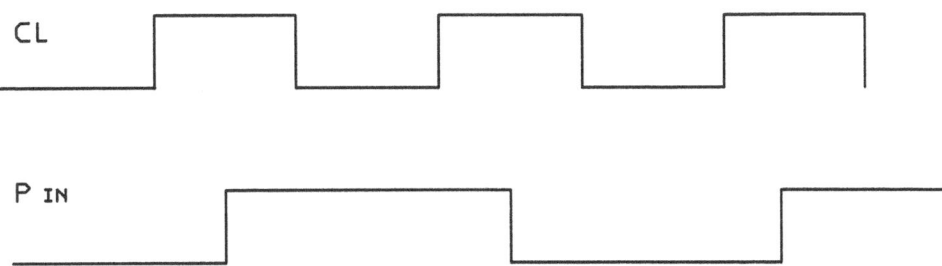

Fig. 3.26. The test pattern of the TrCT for 4014 8-bit shift-register, where the parallel/serial selection input is set to parallel configuration so that each D-type flip-flop can be tested individually.

are plotted together with the exercise patterns in Figure 3.27. The pulse widths and heights are quite different from each other. The widest and possibly the highest pulses are those associated with both outputs changing state. They clearly show the split into two peaks or even more in a pulse. This is mainly due to the delay which occurred at the NOR gate of the slave section which drives the second output buffer. Pulse 1 is only produced by charging node B of Figure 3.17 and therefore is a very narrow pulse. Pulse 5 is produced by the discharging at the gates of the transistors which are connected to the RESET input. It is not associated with the charging or discharging of the internal nodes. We may have noticed that there exist negative peaks at some pulses. They are again associated with the discharging at the gate electrodes of the transistors through the earthed

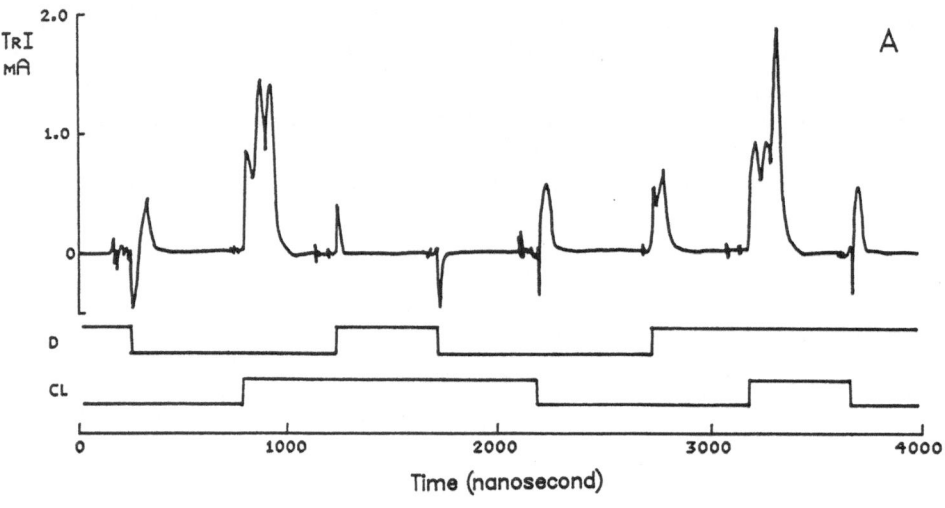

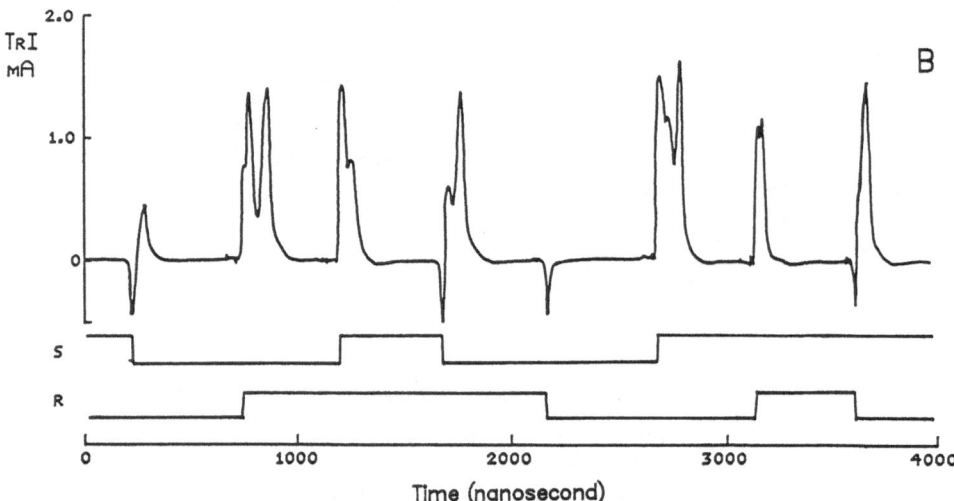

Fig. 3.27. The transient circuit waveform of a D-type flip-flop in the TrCt:(A) toggle mode (B) set/reset mode.

inputs. This discharging process causes the current to flow from the external circuit to the negative supply rail V_{ss}, and thus the current monitored at the negative supply rail will show a negative value. This explains the negative peaks at Pulse 1, 4, 5 and 8 since all these pulses are associated with the high to low transition of the inputs.

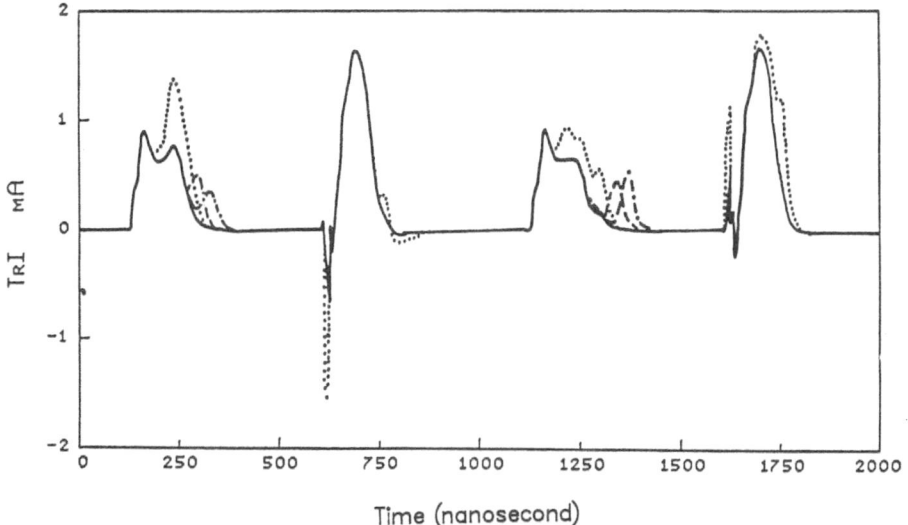

Fig. 3.28. The transient current waveform of a 4014 shift-register in five operating configurations of the TrCT (Curve 1: Stages 1 to 5 exercised; Curve 2 Stage 1; Curves 3,4,5 : Stages 6,7,8, respectively).

The transient current waveforms of the 4014 produced by the test pattern shown in Figure 3.26 are plotted for five predefined measurements of a specific device in Figure 3.28. A total of four pulses are generated because the logic transition of the parallel-in inputs do not produce current pulses when the clock input is held at logic-1. The two higher but narrower pulses are produced by the low to high transition of the clock input, which in turn makes the master section change state. The two wider and shorter pulses are produced when the slave section changes its state, which in turn causes the buffer at the output stage to change state. The tails appearing at some of these pulses are produced by the logic transition of such buffers.

3.3.4. The Static Current Noise Test

3.3.4.1. Introduction to the ScNT

Electrical noise means the voltage or current fluctuation about its mean at the terminals of a component. Note that the fluctuation originates from the random microscopic behaviour of the charge carriers within the components rather that from external interference.

All the components in electronic devices show noise as defined above. Thermal noise in resistors, shot noise in diodes, burst noise in bipolar transistors and flicker noise in MOS transistors are very common noise sources. To be specific to the test vehicle we used for the work, we give below a very brief account of noise sources associated with CMOS technology.

The major thermal noise source in silicon CMOS is the thermal noise of the resistance of the channel of the transistor. The spectral density of the current noise can be characterized as

$$S_i\left(f\right) = S4kT\,g_{do},\tag{3.7}$$

with $1 > S > 2/3$ for a low conductivity substrate and

$$g_{do} = \frac{Z}{L}\mu C_{ox}\left(V_{gs} - V_{Th}\right)\tag{3.8}$$

is the conductance of the channel at small drain voltage.

The observable shot noise in CMOS arises mainly from the leakage current in the reverse biased p–n junction and it is the dominant noise source in the device when the circuit is not switching. The spectral density of the current noise at low frequency is

$$S_i\left(f\right) = 2eI_s,\tag{3.9}$$

where I_s is the 'saturation' current of the p–n junction under reverse bias conditions.

Flicker noise is a resistivity fluctuation with a spectral density given by

$$S_R\left(f\right) = R^2\,\frac{C}{f},$$

where C is a constant which decreases as the amount of the fluctuating species increases. Thus for noise sources distributed throughout a volume, C is inversely proportional to the active volume and for surface noise sources it is inversely proportional to the surface area.

Although many universal theories for the effect exist, the actual source is in many devices not well established. However there is considerable evidence that as the noise increases any device becomes of lower quality, is damaged or is less homogeneous. For MOS devices the noise increases as the surface state density increases. For metallic interconnects the noise increases as electromigration failure progresses. For reverse biased p–n junctions the noise increases as the current increases due to generation–recombination processes via gap states. These are often located at the silicon–silicon oxide interface where the junction depletion region intersects the surface. In general, therefore, the flicker, or $1/f$ noise, is a measure of the quality or defect density of a device. The noise is usually only detectable above the white thermal and shot noise at low audio frequencies. It is revealed as a voltage or current noise when a bias is applied to the fluctuating resistance. In practice the MOS channels exhibit a large flicker noise even if they have been well made.

3.3.4.2. Features of the ScNT

The static current noise test (ScNT) measures the noise in the supply current when the circuit works in the static operation mode. The test system has the form shown schematically in Figure 3.29. It is a sophisticated system since the static supply current of the CMOS circuit is very small. As an example we list some properties of a normal 4013 D-type flip-flop associated with the ScNT in Table 3.11, where we have supposed that only the shot noise of the normal static supply current contributes to the noise measured.

An ordinary current-to-voltage converter will not meet the demand for testing such low-level noise. Instead, a high performance integrator is needed to give low noise current-to-voltage conversion. The noise current is then preamplified and passed through a differentiator to recover the original spectrum. Lastly, the signal is postamplified and is analyzed by the spectrum analyzer. As is shown in the diagram, we have also a part of the circuit which is called the DC bias circuit. This is used to stabilize the DC output of the integrator since the feedback of the integrator is purely a capacitor network. To give a high performance, this part of the circuit consists of a photo-coupler in a circuit with a very long time constant. It provides the DC current bias of the circuit and has shot noise but no flicker noise.

Table 3.11. Electrical Properties of a Normal D-type Flip-flop

Static supply current at 10 A	1×10^{-10} A
Shot noise of supply current	3×10^{-29} A^2/Hz
Impedence $\frac{dV_{dd}}{dI_{dd}}$ at high voltage	$> 1 \times 10^{11}$ Ω

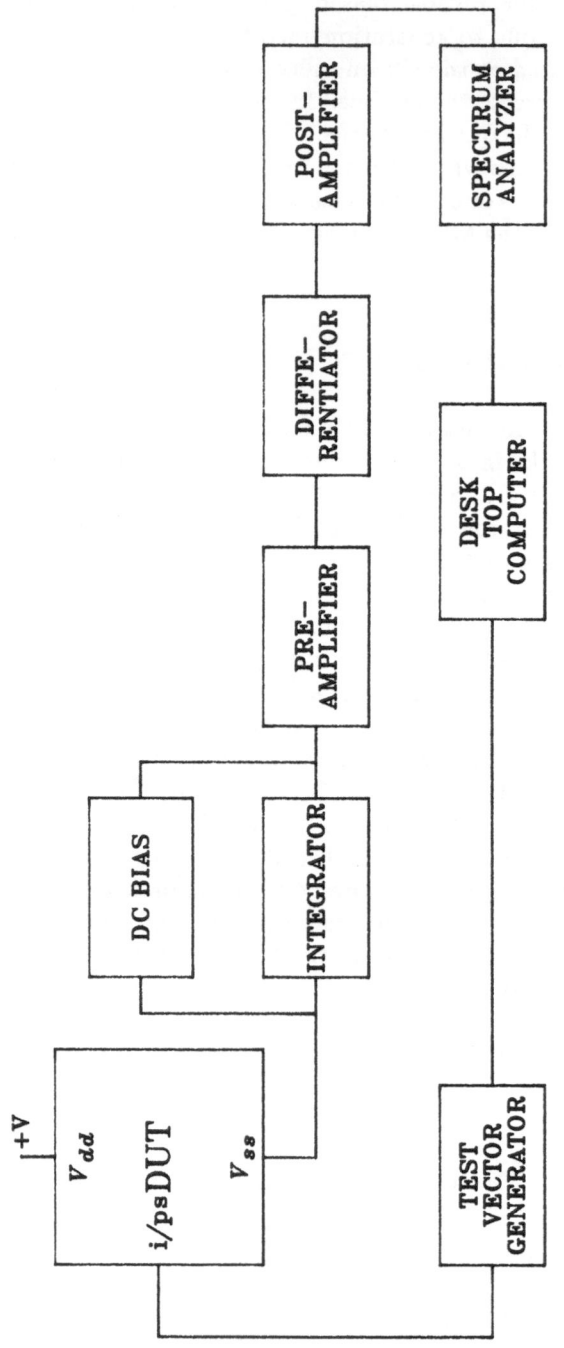

Fig. 3.29. The block diagram of the StNT system.

In addition a pattern generator is used to preset the circuit to some specific state. This is very similar to the StCT since both tests are studying the same parameter which is the static supply current.

A typical spectrum of this type of noise for a normal 4013 D-type flip-flop is plotted in Figure 3.30 together with the shot noise level which is obtained from the equation $S_i(f) = 2eI_s$. The spectrum is flat up to the kiloHertz range and the noise level is quite consistant with the expected shot noise level.

3.3.5. The Transient Current Noise Test

The transient current noise test (TcNT) measures the transient current from the power supply when the circuit works in the dynamic mode. As has been indicated previously, the transient current is composed of the channel current of the transistors which charges or discharges the node capacitances. The TcNT detects the noise in the drain currents of the transistors and hence the resistance fluctuations in the channels. Basically, the test looks for the fluctuation of the amplitude of the current pulse. The implementation of the test uses the system shown schematically in Figure 3.31. It is very similar to the TrCT discussed earlier, since both tests detect the transient current.

Repetitive transient current pulses are first produced by a pattern generator which, for simplicity, may be the same as that for the TrCT. The transient current is then gated by a linear gate to perform a 'sample-and-hold' function. The gate time is controlled by a delay generator. The triggering of the delay generator is provided by the pattern generator to synchronize the gating with the transient pulse. The sampled signed is then averaged by a hold circuit and finally analyzed by the spectrum analyzer.

The transient current for a single pulse during a transition lasts from a few tens to hundreds of nanoseconds. A 10 nanosecond width gate window, which is about several tenths of the transient current width, is chosen together with a 10 nanosecond interval between the successive gating positions to get a detailed analysis of the entire transient current waveform with a reasonable resolution and without taking too much time. The gate delay time is slowly increased to scan the gate across the pulse.

This test has to be fully automated since the time needed to perform a measurement for a single transient waveform can be as long as an hour and the precise control of the analog signal can be very difficult if it is manually adjusted. First of all a delayed gate window is produced in the scan delay generator. The amount of delay is determined by a D to A converter. Second, the transient current signal is gated and averaged during the gate signal of the scan delay generator. Third, the analogue output of the 415 linear gate is fed into a digital voltmeter (DVM) and the HP3582A spectrum analyzer at the same time. Lastly, the data for both signal amplitude and noise measurement are stored. When this cycle finishes, a further delay of the gate window in the scan delay generator is produced by the D to A converter and a new cycle starts. When the entire

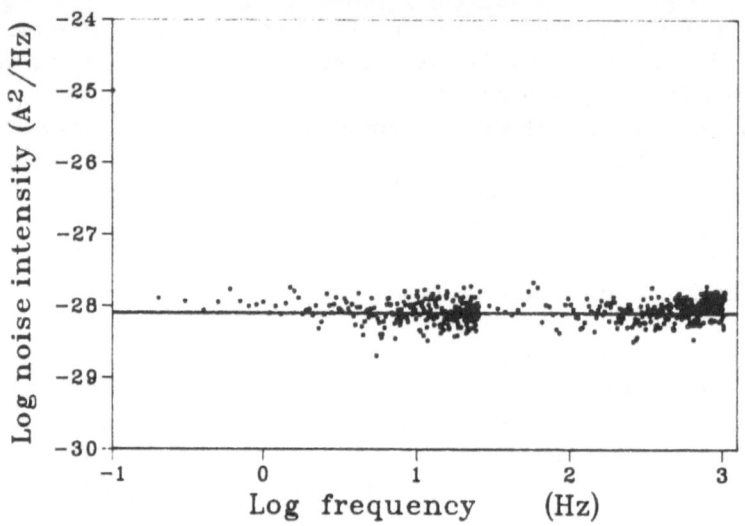

Fig. 3.30. The low frequency noise spectrum in the static supply current of a normal D-type flip-flop.

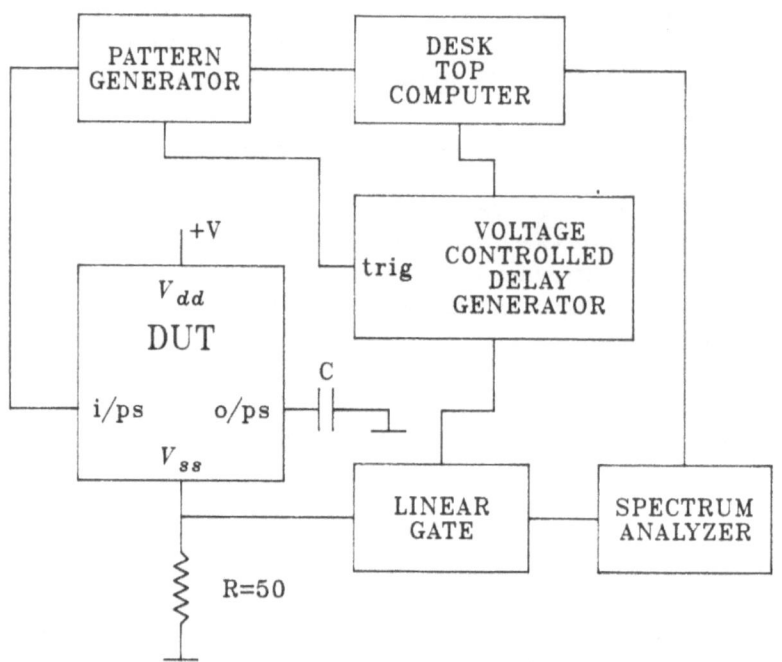

Fig. 3.31. The block diagram of the TrNT system.

transient peak is scanned the measurement is over. By processing the data stored, the original waveform is recovered by calibrating the data obtained from the DVM, and the low frequency noise properties of the entire waveform is produced by calibrating the data from the HP3582A. Hence both the current waveform and the related noise on it can be drawn on the same graph.

In our system it has been found that the noise of the device is usually masked by the noise of the test system unless the circuit becomes defective.

3.3.6. The Threshold Voltage Test

The threshold voltage and mobility of the channel carriers are key parameters in MOS technology. This is due to the fact that the basic feature of the operation of an MOS circuit is the charging and discharging of node capacitances by the available channel current of the transistors. The tests devised in the previous subsections are very much associated with these two parameters. In this subsection we present a test which gives a direct measure of the threshold voltages and mobility of channel carriers in an MOS transistor.

A well defined linear dependence of the square root of the drain current on gate voltage holds when the MOS transistor is biased in a saturation region with $V_{dd} \gg |V_{gs} - V_{Th}|$. It is then possible to calculate the gain factor and the threshold voltage if a voltage scan at the gate is performed, as long as the device is still working in the saturation region.

In a CMOS inverter, at first sight, we cannot access each transistor in the pair to perform the above measurement. However, in fact we can characterize each transistor in the pair to get the gain factors and threshold voltages. Let us consider an inverter in switching operation. When V_{in} is held at 0 V the p-channel transistor is fully ON and the n-channel is fully OFF. The output voltage is the positive supply rail V_{dd}. As V_{in} is increased to exceed the threshold voltage of the n-channel transistor V_{Thn}, the n-channel transistor gradually turns on, and in this region the supply current is usually limited by the drain current of the n-channel transistor, which has the form:

$$I_{dd} = K_n [V_{in} - V_{Thn}]^2. \tag{3.10}$$

A further increase of V_{in} results in a dramatic increase in the drain current of the n-channel transistor. The output voltage decreases dramatically. After the above transition region the opposite happens with the n-channel transistor fully ON and p-channel transistor OFF. In this region the supply current is limited by the channel current of the p-channel transistor:

$$I_{dd} = K_p [V_{in} - (V_{dd} - |V_{Thp}|)]^2. \tag{3.11}$$

Therefore, the expected supply current versus the gate voltage associated with a CMOS inverter should show the variation depicted in Figure 3.32. A maximum value of the supply current appears when the channel currents of both transistors are equal.

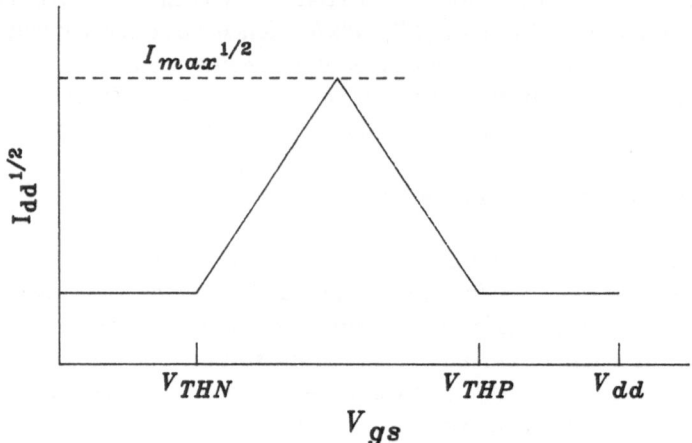

Fig. 3.32.   The characteristics of the square root of the supply current versus gate voltage in a basic inverter.

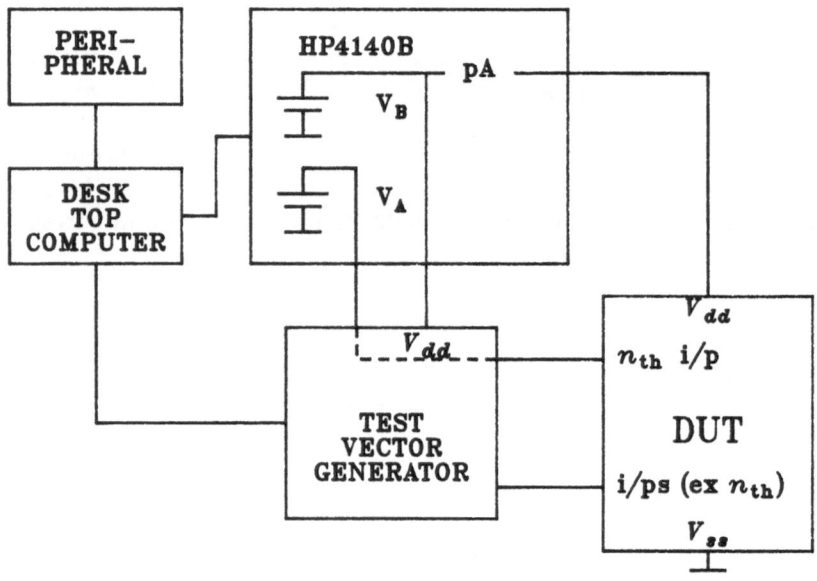

Fig. 3.33.   The block diagram of the ThVT system.

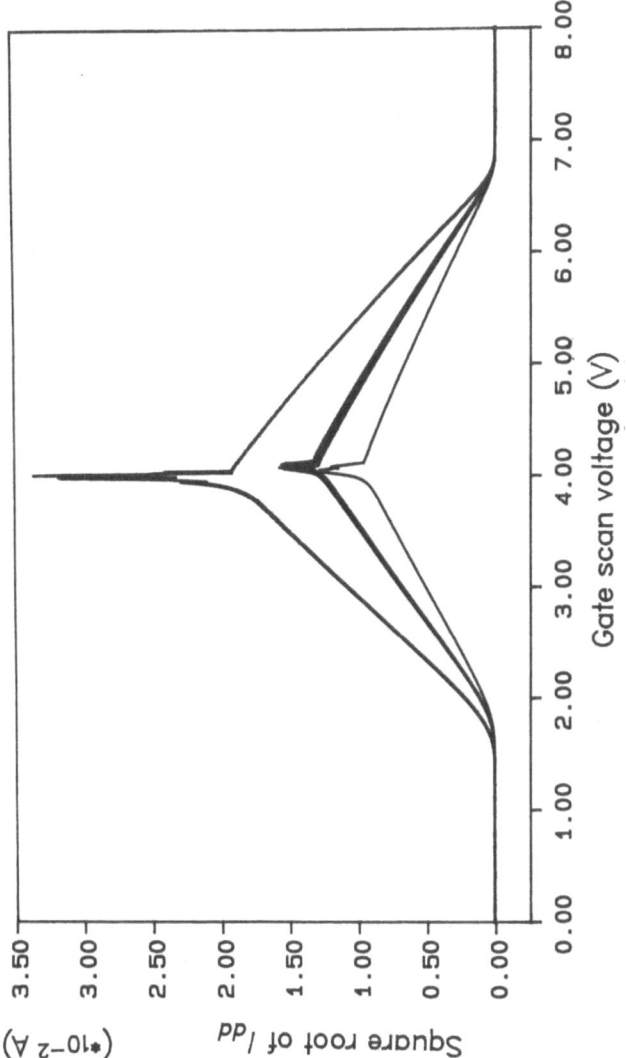

Fig. 3.34. The characteristics of the square root of the supply current versus scan voltages applied to different input transistor pairs, where 1 ∼ 11 corresponds to the eleven inputs listed in Table 3.12.

We can get both threshold voltage and gain factor of the transistors by using Equations 3.10 and 3.11 in the $V_{Thn} < V_{in} < V_{out}$ and $V_{out} < V_{in} < (V_{dd} - |V_{Thp}|)$ regions, respectively.

In the threshold voltage test (ThVT) we measure the total supply current as the voltage to an input terminal is scanned from negative to positive rails and then fit the relevant data into Equations 3.10 and 3.11 to get the values of threshold voltages and gain factors for both transistors in the pair.

Table 3.12. The Values of the Threshold Voltage and the Gain Factors Obtained from the Curve Fitting on Measured Data for the Circuit Shown in Fig. 2.2(b)

		V_{Thn}	$K_n \times 10^{-5}$	V_{Thp}	$K_p \times 10^{-5}$
P/S		1.88	7.38	1.15	5.33
CL		1.87	7.3	1.11	5.75
Sin		1.89	1.95	1.08	1.38
Pin	1	1.87	3.25	1.10	2.64
	2	1.87	3.28	1.10	2.53
	3	1.86	3.23	1.10	2.62
	4	1.85	3.27	1.10	2.74
	5	1.87	3.32	1.06	2.59
	6	1.87	3.28	1.09	2.63
	7	1.86	3.17	1.10	2.60
	8	1.85	3.33	1.06	2.57

A block diagram of the ThVT is shown in Figure 3.33. The DUT is biased to a predefined state and the nth input which is to be tested is connected to the scan voltage source V_A. The supply current is measured with each increment of the scan voltage. The test of each input inverter is done through demultiplexing V_A onto each input terminal and biasing the rest of this to a predefined state. A typical feature of the supply current against the scan voltage V_{in} is plotted in Figure 3.34 for a normal 4014 shift register, where a total of 11 curves are shown for 11 inputs. Before discussing these curves, we will first examine the results obtained from the curve fitting which are listed in Table 3.12.

As listed, the threshold voltages of both n-channel and p-channel transistors are quite consistant among all the inputs. The deviation of the values of the gain factors for both p-channel and n-channel transistors arises mainly from the dimension of the transistors. By taking the dimensions of the transistors into account, we can also find a ratio of 2:1 existing between the mobility of the n-channel transistor and that of the p-channel transistor. Now we account for the curves shown in Figure 3.34. Firstly, the turn-on voltages at both corners are very similar for all curves. This indicates the consistency of threshold voltages listed in Table 3.12. The deviation of the slopes of the curves arises from the difference betweeen the device dimensions. At the top of the experimental curve various sharp peaks are observed. These are due to the currents passed by the various internal gates as they switch when their input voltage reaches their threshold.

3.4. ORGANIZATION OF THE EXPERIMENT

In the previous sections a detailed discussion has been given of the individual elements of the experiment such as the tests developed, the stress system used and the specimens used in this project. In this section an overall account of the experimental organization is presented.

The test program is organized in such a way that it studies the changes of the parameters and of the noise level and characteristics of the devices being studied. The prestress measurements for all tests were taken before anything was done to the specimens. Two sets of the prestress measurements were done for the tests developed in this project so that they could be compared and precise values obtained to serve as a good reference level. The specimens were then stressed for a time under the specified conditions to accelerate the degradation and failure of the specimens. Then after being allowed to cool down for a certain period of time to room temperature, the stressed specimens were sampled using a specific procedure and those specimens chosen by this sampling procedure underwent the StCT, CoFT, TrCT and SCNT. The optional and time consuming TCNT was performed only for those devices which behaved abnormally during the TrCT check. The results of the tests were assessed by comparing them with the results of the prestress tests. Finally, on the basis of the above assessment, some specimens might be taken away from further stresses and be substituted by new devices. We call such a 'loop' program from stressing, testing, assessment to stress device selection for the next stress a 'RUN', a term which will be widely used in the remaining part of this book.

The purpose of the prestress tests is mainly to eliminate the very poor devices and to set a reference for the comparison of the poststress tests.

The testing program for Batch 1 was started as a trial. It was gradually improved as testing progressed. For this reason some of the measurements made and the sequence of the test may not be ideal. Initially this was only considered a rough experiment, but the results were interesting enough to warrant subjecting the batch to a full investigation.

The stress process was arranged to allow the stressed devices to be properly annealed. After the power of the oven was switched off, the door of the oven was not opened and the electrical stress not stopped until the specimens had cooled down. Normally such a system was left to run overnight, or roughly 12 hours, after the oven power had been turned off.

A relaxation time was found to be needed for the stressed devices to recover themselves to an equilibrium state before the poststress tests could be done. The recovery shows itself as a slow change of the test parameters even when the device is at an equilibrium temperaure. The processes involved are not known but probably involve ion migration. We have found the time constant of this process experimentally to be over 8 hours.

After the specimens had been stressed, annealed and relaxed, the poststress tests were done on the sampled specimens chosen according to the following rules: (1) the specimens appearing abnormal in the stress; (2) the specimens which passed the functional tests during the stress but showed abnormal changes in the poststress tests of the previous run; and (3) preselected normal specimens that are tested for a routine check. The number of the samples to be tested in each run was chosen to be ten or eleven. For Batch 1, five selected devices were sampled every run to allow for the most frequent monitoring of the parameters and hence the most detailed analysis of the properties of the parameters during the successive runs.

The period of each stress was chosen by experience so that there would be a reasonable expectation of some change in the batch during the stress period. Initially the stress period chosen was about a day so that early changes would be observed, but finally periods of over 10 days were studied.

REFERENCES

General

Gottlieb, G.E., 1982, *Tech. Digest IEEE Test Conference*, 287-98.
Miller, D.M., 1988, "Developments in Integrated Circuit Testing", Academic Press, London.
Reynolds, F.H., 1974, *Proc. IEEE*, **62**: 212-22.
Stitch, M., Johnson, G.M., Kirk, B.P., and Brauer, J.B., 1975, *IEEE Trans. Reliability*, **R-24** : 238-50.
Wilkins, B.R., 1987, "Testing Digital Circuits", van Nostrand Reinhold.

Digital Fault Testing

Al-Arian, S.A., and Agrawal, D.P., 1987, *IEEE Trans. Circ. Syst.*, **CAS-34**: 269-79.

Banerjee, P., and Abraham, J.A., 1984, *IEEE Design and Test*, 76-86.

Galiay, J., Crouzet, Y., and Vergniault, M., 1980, *IEEE Trans. Computers*, **C-29**: 527-31.

Jain, S.K., and Agrawal, V.D., 1985, *IEEE Trans. Computers*, **C-34**: 426-33.

Kodandapani, K.L., and Pradhan, D.K., 1986, *IEEE Trans. Computers*, **C-29**: 55-9.

Malaiya, Y.K., 1984, *IEEE Conf. on Computer-aided Design*, 248-50.

Mallela, S., and Wu, S., 1984, *Tech. Dig. Int. Test Conf.*, 57-61.

Moritz, P.S., and Thorsen, L.M., 1986, *IEEE J. Solid-state Circuits*, **SC-21**: 306-9.

Nickel, V.V., 1980, *Tech. Digest IEEE Test Conf.*, 378-81.

Wadsack, R.L., 1978, *Bell. Syst. Tech. J.*, **57**: 1449-73.

Static Current Test

Hawkins, C.F., and Soden, J.M., 1985, *Tech. Digest. Int. Test Conf.*, 544-54.

Horning, L.K., Soden, J.M., Fritzemeier, R.R., and Hawkins, C.F., 1987, *Proc. Int. Test Conf.*, 300-9.

Jacomino, M., Rainard, J.L., and David, R., 1987, *Proc. Third Int. Conf. GI/ITG/GMA*, 83-94.

Levi, M.W., 1981, *Tech. Digest. IEEE Test Conference*, 217-20.

Malaiya, Y.K., 1985, *Microelectron. Reliab.*, **25**: 943-8.

Malaiya, Y.K., and Su, S.Y.H., 1983, *Tech. Digest IEEE Test Conf.*, 25-34.

Soden, J.M., and Hawkins, C.F., 1987, *Semicond. Int.*, May: 240-5.

Stojadinovic, N.D., 1983, *Microelectron. Reliab.*, **23**: 609-707.

Cut-off Frequency Test

Ager, D.J., Cornwell, G.F., and Stanley, I.W., 1982, *Microelectron. and Reliab.*, **22**: 241-64.

Ager, D.J., Cornwell, G.F., and Stephens, C.E., 1982, *Reliability Engineering*, **3**: 413-22.

Chaudramouli, R., and Sucar, H., 1985, *Tech. Digest Int. Test Conf.*, 313-21.

Koeppe, S., 1986, *Tech. Digest Int. Test Conf.*, 530-6.

Martinez, A.M., 1989, *IEEE J. Solid-state Circuits*, **SC-24**: 520-31.

Moltoft, J., 1983, *Electrocomponent Science and Technology*, **11**: 71-84.

Shimono, T., Oozeki, K., Takahashi, M., Kawa, M., and Funatsu, S., 1985, *Tech. Digest Int. Conf.*, 329-49.

Wagner, K.D., 1985, *Tech Digest Int. Test Conf.*, 334-41.

Noise Tests

Chen, T.M., Djeu, M.P., and Moore, R.D., 1985, *IEEE Int. Reliab. Phys. Symp.*, 87-92.

Diligenti, A., 1985, *IEEE Electron. Devices Lett.*, **EDL-6**: 606-8.

Gupta, M.S., 1975, *Proc. IEEE*, **63**: 996-1010.

Herman, E.M., 1983, *J. Appl. Phys.*, **54**: 1937-49.

Koch, R.H., Lloyd, J.R., and Cronin, J., 1985, *Phys. Rev. Lett.*, **55**: 2487-90.

Maes, H.E., and Usmani, S.H., 1983, *J. Appl. Phys.*, **54**: 1937-49.

Neri, B., Diligenti, A., and Bagnoli, P.E., 1987, *IEEE Trans. Electron. Devices*, **ED-34**: 2317-22.

Peczalski, A., 1983, *IEEE Computer Society Curriculum for Test Technology*, 37-40.

Pimbley, J.M., 1984, *IEEE Electron. Devices Lett.*, **EDL-5**: 345-7.

Stocker, J.D., 1984, *PhD Thesis*, University of Lancaster.

Threshold Voltage Test and Radiation Effects

Lim, T.S., Martin, R.L., and Hughes, H.L., 1986, *IEEE Circuits and Systems*, Sept: 29-33.

Lim, T.S., Martin, R.L., and Hughes, H.L., 1987, *IEEE Circuits and Systems*, January: 22-30.

Prince, J.L., Draper, B.L., Rapp, E.A., Kronberg, J.N., and Fitch, L.T., 1980, *IEEE Trans. Components, Hybrids Manuf. Tech.*, **CHMT-3**: 135-44.

Sah, C.T., 1976, *IEEE Trans. on Nucl. Sci.*, **NS-23**: 1563-8.

Stojadinovic, N., Dimitrijev, S., Mijalkovic, S., and Zivic, Z., 1983, *Phys. Stat. Solidii*, (a)**76**: 357-64.

Thompson, J., Rogers, T., and Galey, R.A., 1983, *Proc. Inst. Radio and Electronic Eng.*(Australia), September: 117-9.

Various authors: *Proceedings of the 1986 Annual Conference on Nuclear and Space Radiation Effects, 1986, in IEEE Trans. Nucl. Sci.*, **NS-33**: 1161-718.

Chapter 4

ASSESSMENT OF THE TESTS AS
PREDICTORS OF FAILURE

4.1. BEHAVIOR OF THE DEVICES SUBJECT TO THERMAL AND ELECTRICAL STRESS

4.1.1. Reference to Experimental Details

In this section we will illustrate the usefulness of the devised tests by referring to the behavior of the devices subject to thermal and electrical stress. Before we discuss this in detail in the next few subsections we first cf all introduce the experimental details for reference.

A summary of the batches which were used in thermally and electrically accelerated life testing is given in Table 4.1.

A total of three batches of the CMOS circuits have been studied in this program. Two of them are 4013 D-type flip-flops and the third one is a 4014 8-bit shift-register. All 77 chips are identified by the notation listed in Table 4.2. In addition Batch 3, which is missing in these tables, was studied under radiation stress and will be discussed in the next section.

Table 4.1. The Experimental Details of the Specimens Subjected
to Thermal and Electrical Stress

Batch number	Device type	Number of samples	Stress conditions	Total of runs/days
1	4013 Plastic-packages	27	182°C, 5 V Dynamic(1 kHz)	48/256
2	4013 Ceramic-packages	25	220°C, 15 V Dynamic(1 kHz)	36/294
4	4014 Ceramic-package	25	220°C, 15 V Dynamic(1 kHz)	11/64

Table 4.2. Reference Numbers of the Devices Studied Under Thermal and Electrical Stress

Batch	Device idendification	
1	B1 4013 XX/1+2*	XX=01 → 09, 26 → 30, 36 → 44, 51 → 52
2	B2 4013 XX/1+2	XX=11 → 35
4	B4 4014 XX	XX=01 → 25

* '1+2' Distinguishes one device from the other in a single chip with device '1' connected to Pins 9 to 13.

Some notation may need explanation as it will be widely used in the following sections. A 'RUN' is defined as a cycle which includes stress and poststress test. The RUN number appears in a form such as 'Run 02/3' with the first number '02' indicating the number in the sequence of the stress-test runs with zero as the prestress test, and the second number '3' indicating the total number of days that the particular specimen has been exposed to the stress at the time of the test. The word *degradation* will be used to indicate different effects depending on the context in which it is used. It may be narrowly used to indicate excess changes of the parameters in the devised tests but may not necessarily indicate the degradation of the device performance. In addition, the 'TOGGLE' 'SET' and 'RESET' are used in the CoFT to indicate the specific operation modes of the device being tested, and 'TOGGLE' and 'SET/RESET' are used in the TrCT with the same meaning as in CoFT. These terms were defined in Section 3.3.

4.1.2. Discussion of the StCT

Many types of defects in the CMOS circuit give rise to excess supply current in the unexercised operation condition. These types of defects usually cause a shift of the gate-oxide and field-oxide threshold voltage, short or open circuit of the tracks, degradation of the p–n junctions, and so on, which have been discussed briefly in Chapter 3. As also has been discussed there, the current produced by the inherent defects in well made devices is usually less competitive compared with the generation current of the reverse biased p–n junction. The excess current which dominates the supply current occurs only when abnormal defects exist in the circuit. Such defects may be divided into two classes: the latent defects that may not appear while being tested but may appear later in operation, and the abnormal defects that result in excess current while being tested. In this subsection we talk mainly about the latent defects. The abnormal defects that appear inherently in the purchased new devices are discussed in Section 4.3. Note that so called latent defects and abnormal defects in the context of this subsection are referred to in the StCT depending on the observability of the excess supply current and therefore have different meanings when compared with the latent defects which are generally used in the book and which are referred to as the defects that are observable in some of the tests derived in this book but do not cause functional failure of the circuit.

To illustrate these points, we present in the following paragraphs the general features of the StCT obtained from our work. After having been stressed, normal specimens did not show significant changes in the static supply current for a very long period of time. The general features of the variation of the current with stress time for two normal specimens is shown in Figure 4.1. The static supply current did not vary more than 10 % after 256×24 hours of stress. The random fluctuation from test to test, especially in the beginning of the experiment for Batch 1, resulted from the temperature fluctuation of the testing environment since the generation current of a p–n junction varies very rapidly with the temperature. This feature has also been observed in Batch 4 which exhibited a few percent variation in the static supply current.

Batch 2 showed, however, different characteristics of variation in the static supply current with stress. A general increase of the current with stress time, especially in the beginning of the stress, has been observed for all normal devices in the batch. The results for two normal specimens are shown in Figure 4.2. The specimens such as B2 4013 27/1+2 showing a large static supply current in the prestress test are affected less by the accelerated stress than are specimens such as B2 4013 30 which show a small current in the prestress test. The difference in the static supply current from specimen to specimen became less and less and the trends of the increase gradually faded away with the progress of the stress. This effect could probably be contributed by a burn-in effect on the semiconductor devices, although such a long-term burn-in effect needs questioning.

The above results indicate that, in principle, the accelerated stress employed in the work introduces unexpected failures such as those caused by the package materials. However there are devices which are affected by the stress because of the latent defects in the device.

Figures 4.3 displays the results of the StCT for sample B1 4013 26/1+2 in Run 26/48. The supply current was measured over the 20 vectors that were listed in Table 3.8 for both devices on the chip. The excess current stands significantly above the intrinsic supply current in Figure 4.3 and this current only appears at some test vectors. This is expected because the excess current can normally be stimulated only by some of the test vectors, although the global appearance of the excess current over the entire test set is possible in some very rare cases. In addition, Device 2 on the chip displayed no observable changes over the entire test set as shown in Figure 4.3. Care is needed when we consider the test set for two devices on a single chip as two separate test sets. That is, one device has to be set to a specific test vector at which there is no excess current appearing while the other device on the same chip is tested. This reveals that some type of defect or defects exists on the part of the circuit where B1 4013 26/1 is located and there is no abnormal defects existing in B1 4013 26/2 as far as the StCT is concerned.

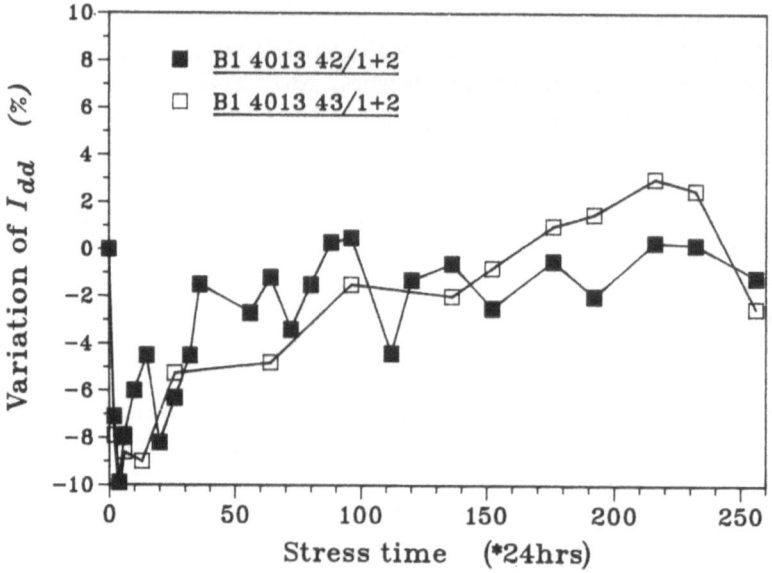

Fig. 4.1. Typical variation of the static supply current with voltage–temperature stress for normal specimens in Batch 1.

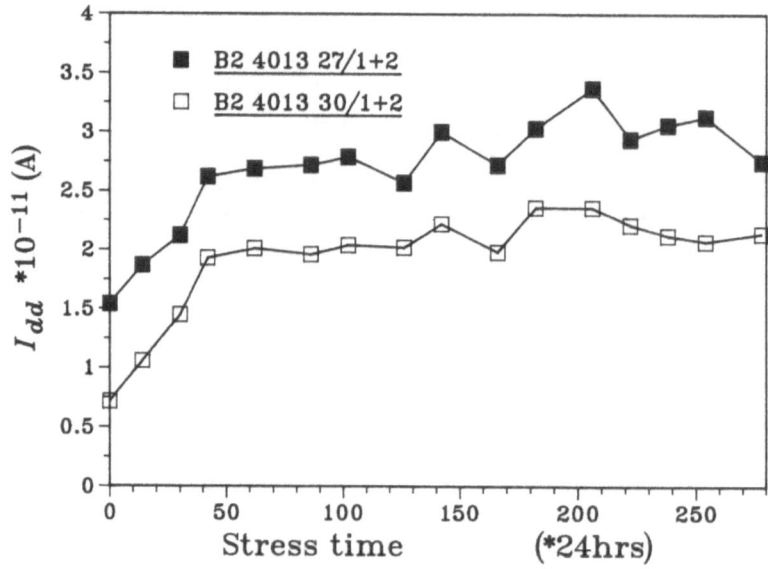

Fig. 4.2. Variation of the static supply current with voltage–temperature stress for normal specimens in Batch 2.

98

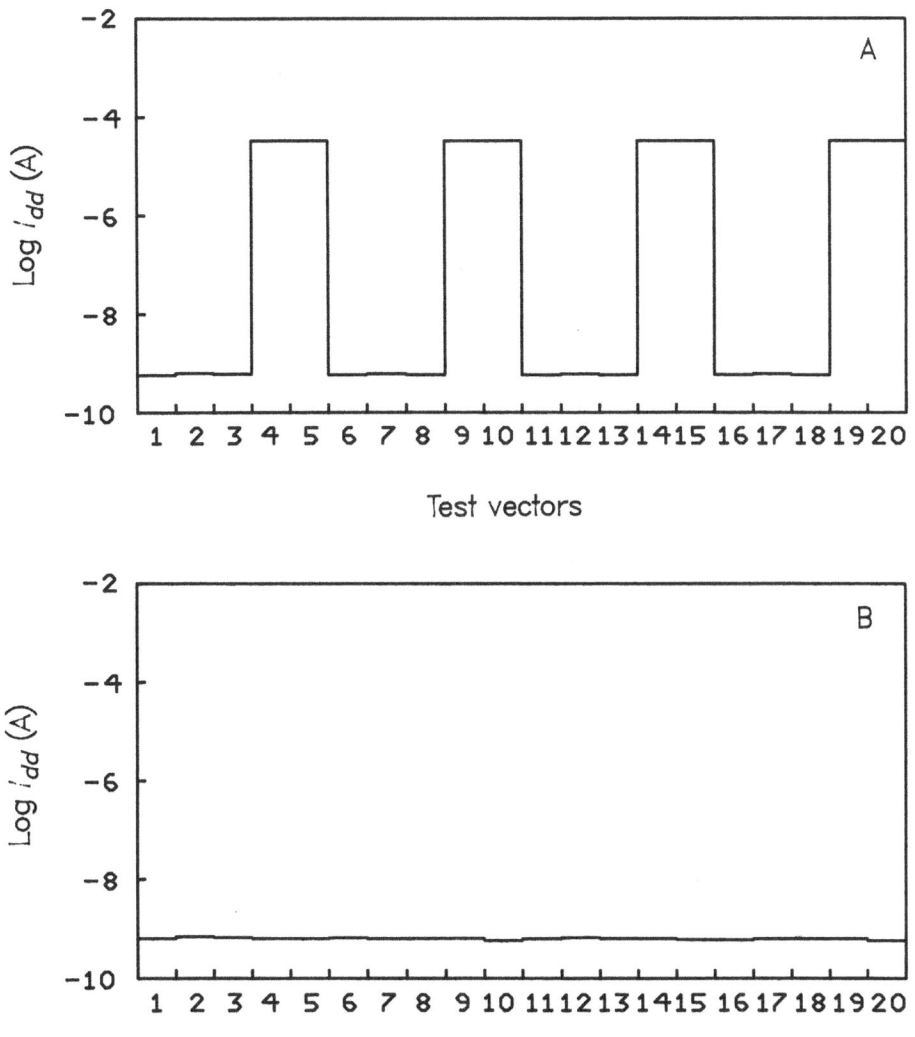

Fig. 4.3. The static supply current against the test vectors of the test set for specific subnormal devices of Batch 1. (A) Device 1 of two devices in a chip B1 4013 26/1+2 is exercised; (B) Device 2 is exercised. In each case the unexercised device is set to vector 01.

If the test is used simply to detect if there are any defects that cause an excess supply current in a circuit then the above procedure is adequate since the test goes through all the possible states of the circuit and therefore any 'stuck-at' type of defects will be revealed. The pattern of the excess supply current as displayed in the above example does not provide, however, all the information on the defect needed for the purpose of the failure analysis. Some questions still remain. What does the pattern of the excess supply current indicate? That is, how many individual defects are associated with this specific pattern of the excess current? What type or types of defects are present? Further analysis of the excess current by doing the detailed $I \sim V$ measurement is necessary to find the answers to these questions. Each type of defect associated with semiconductor devices give rise to a unique $I \sim V$ characteristic. Some typical $I \sim V$ curves associated with the related defects in CMOS technology are illustrated in Figure 4.4. A poor quality interface of the p–n junction could introduce excess current because of surface conduction. This phenomenon usually produces the $I \sim V$ characteristic shown by Curve 2 of Figure 4.4. Defects such as metallic precipitates in the depletion region within the bulk may produce soft $I \sim V$ characteristic shown by Curve 4. It is reported that electrically active stacking faults which intersect a p–n junction at the silicon interface can also contribute to such a soft characteristic. Field oxide thresholds such as excess oxide charges cause an enormous increase of the current at relatively low voltages as shown by Curve 5. The surface breakdown of the p–n junction caused by, for example, mechanical stress or accumulation of charges at the surface will result in the $I \sim V$ characteristic with reduced breakdown voltage as shown by Curve 3. Therefore, the detailed $I \sim V$ measurements over the test vectors greatly helps the interpretation of the current pattern and the defects associated with it. Now, let us examine the experimental results of the above sample. The $I \sim V$ characteristics of B1 4013 26/1 at test vector 01 and 20 are plotted in Figure 4.5. For clarity we plot only two curves since all other curves are the same as either the intrinsic $I \sim V$ curve or as the one at vector 20. By now we may conclude that the excess supply current appearing in B1 4013 26/1 is generated by a defect that causes some type of breakdown since the drastic increase of the current appears only at very high voltages.

A single fault is usually assumed in a circuit to derive the test set to detect the abnormality. We also supposed this condition in deriving the test set for the StCT. Although it is a coincidence, the B1 4013 26/1 really follows the assumption. However, this assumption does not prevent the possibility of multiple defects from occurring in the circuit. A good example of multiple defects is shown in Figure 4.6 which was obtained from B4 4014 25, a 4014 shift-register in Batch 4. The numbers $1 \sim 10$ in the diagram indicate the number of the test vectors at which the $I \sim V$ measurements were done. The curves can be divided into four groups: Curves 1, 8, 9; 4, 5, 10; 2, 3; 6, 7. Curves 6 and 7 which did not show any significant changes display the intrinsic supply current of the circuit. It seems that there are three different defects associated with the appearance of these curves. However, there are only two types of defects existing in this particular

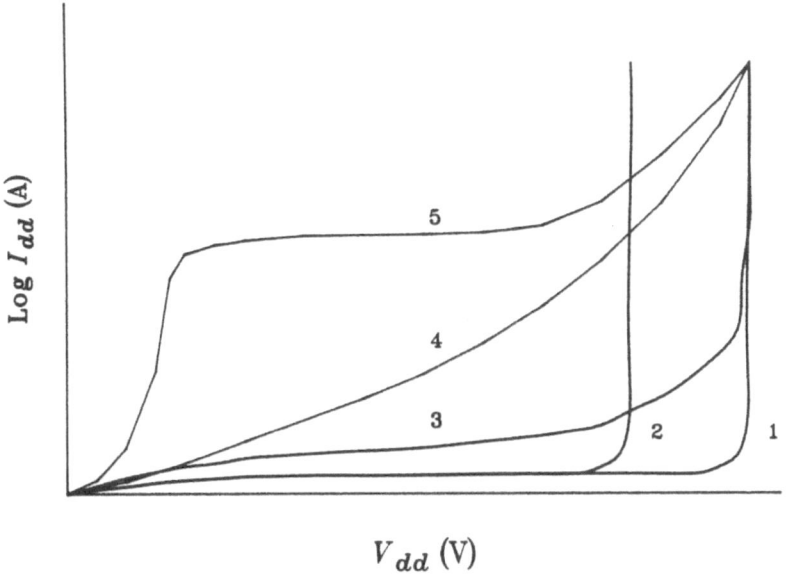

Fig. 4.4. Some typical $I_{dd} \sim V_{dd}$ characteristic curves associated with specific defects in CMOS technology.

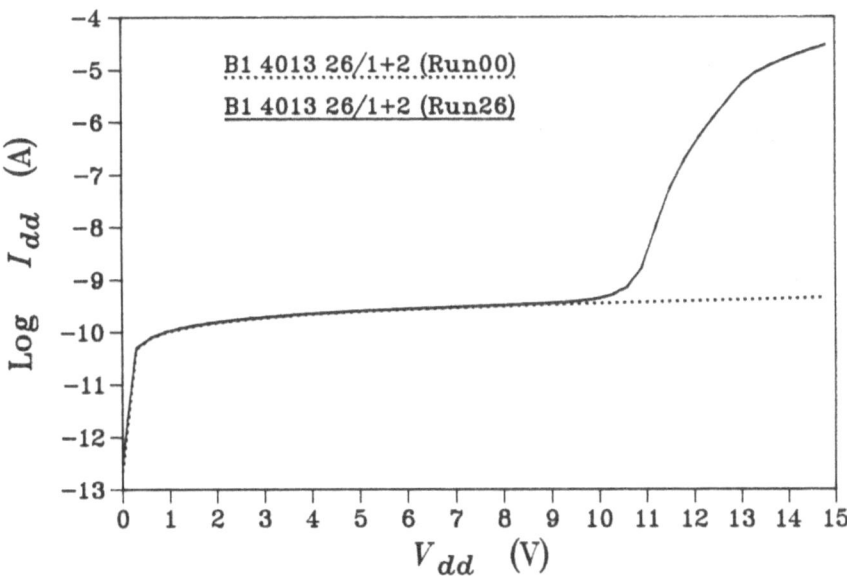

Fig. 4.5. The $I_{dd} \sim V_{dd}$ characteristics of B1 4013 26/1+2 at test vector 01 of Device 1; Device 2 is set to vector 20.

circuit, which we will explain later with Curves 1, 2, 3, 8 and 9 resulting from one type and Curves 1, 4, 5, 8, 9 and 10 from other type. It is clear that the above analysis would be misleading if one only looked at the current values at 15 V and without knowing the detailed $I \sim V$ characteristics since there are three distinctive groups of the curves that are far above the intrinsic current curves.

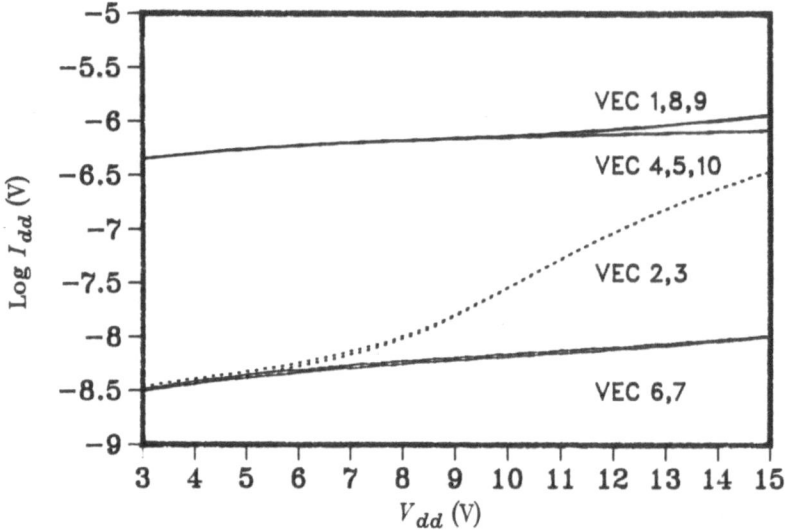

Fig. 4.6. The $I_{dd} \sim V_{dd}$ characteristics of B4 4013 25 at all test vectors of set. This diagram illustrates the multiple faults in a circuit.

The excess supply current in a defective CMOS circuit usually appears a few orders of magnitude higher than that of the intrinsic supply current in the same type of normal device since the amplitude of this intrinsic current is very small, usually not over nanoamperes for SSIs and MSIs, and microamperes for LSIs and VLSIs. The two examples given above show two to five orders of maginitude difference between intrinsic and excess current values, despite the fact that such large differences may be only gradually reached with the progress of the stress and only a small section of the whole circuit appears to be defective. Thus the StCT exhibits very high sensitivities in the detection of the excess current and therefore the application of the test to VLSIs should be feasible.

4.1.3. Discussion of the CoFT

After having been subjected to thermal and electrical stresses, the normal devices do not exhibit significant variations in cut-off frequency. Typical variation in cut-off frequency versus stress time are shown in Figure 4.7(A) and (B). The data were obtained in the toggle mode and at a supply voltage of 3 V. Two plots exhibit the general feature of the CoFT on devices subject to the thermal and electrical stress. The plot of Figure 4.7(A) illustrates a burn-in effect in the beginning of the stress. The cut-off frequency decreased by almost 10 % in the first few runs. After this stage the cut-off frequency stabilizes with almost negligible variation over further stress. The plot of Figure 4.7(B) illustrates an insignificant variation of the cut-off frequency over an entire period of 256×24 hours of stress. It did not even show any burn-in effect in the beginning of the stress. It has been observed that all the specimens in Batch 1 and 4 show similar characteristics to Figure 4.7(B) and all the specimens in Batch 2 show exactly the same effect as Figure 4.7(A). This may suggest that some newly made devices do not go through the burn-in procedures before shipping, although a 10 % decrease of the cut-off frequency may occur after a period of time when such devices go into operation. This situation is undesirable.

Nevertheless, the above results clearly indicate that a normal specimen should not show significant variation in cut-off frequency in a relatively long period of time under the stress conditions mentioned in the beginning of this section. Only devices with latent defects should show degradation in cut-off frequency. We present the general features of some subnormal devices in the following paragraphs.

It has been widely observed that the cut-off frequency at low supply voltage is more sensitive to the stress than that at high voltages. As shown in Figure 4.8 the degradation of the cut-off frequency at 3 V supply voltage with stress is far greater than that at 5 V. This effect is, in a way, explicable if such degradation is caused by mechanisms such as the threshold voltage shift since the drain current will depend closely on the threshold voltages when this is comparable to the power supply voltage. Such interpretation only accounts for some specific type of degradation. The degradation of the cut-off frequency may result from different mechanisms such as excess leakage current appearing at the active nodes. If such current only appears at high voltages we then reach the conclusion that the degradation at high voltages is more than that of low voltages. However, it seems that the experimental results obtained so far show the former feature. Hence the results presented below will be only for a supply voltage of 3 V.

The general feature of the CoFT is that a normal device shows little significant change in cut-off frequency with stress. This, in turn, indicates that an abnormal change of this parameter indicates the degradation or failure of the device. To illustrate this the results from a specific device are plotted as the variation of its cut-off frequency versus stress time in Figure 4.9. Despite the scatter of the data, it is apparent that the cut-off frequency decreases gradually

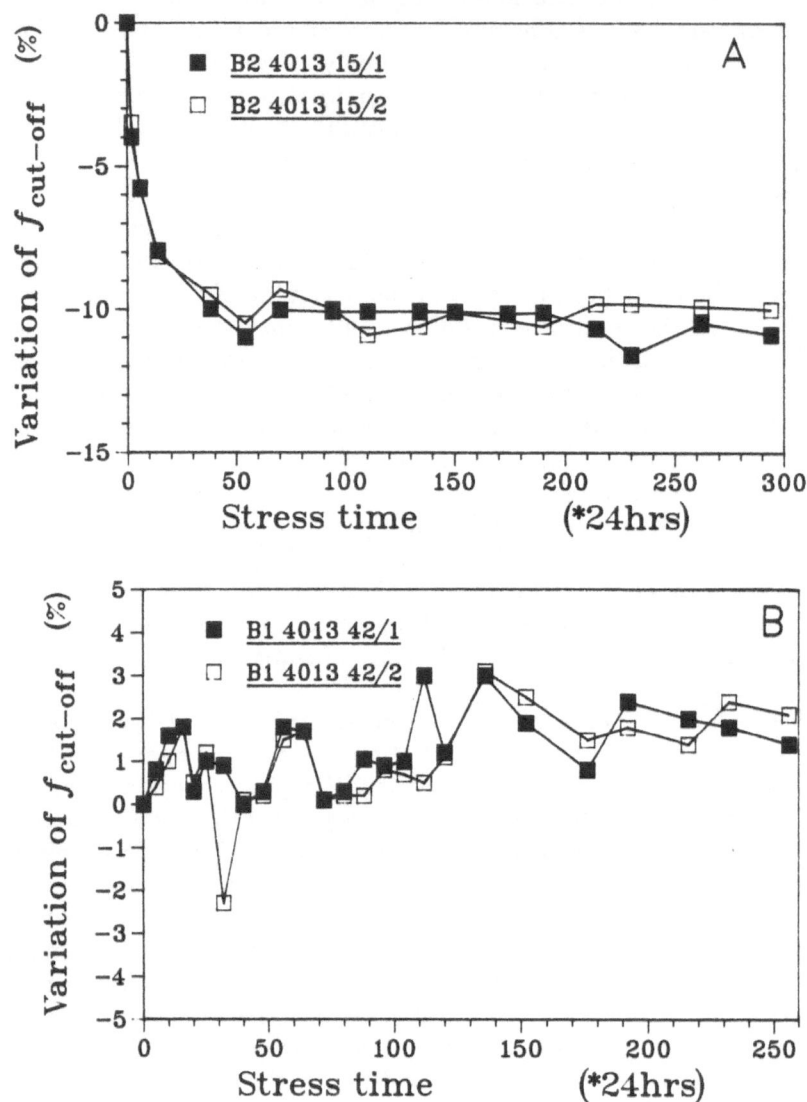

Fig. 4.7. Variation of the cut-off frequency with voltage–temperature stress for normal specimens: (A) showing the characteristic of Batch 2 and (B) showing the characteristic of Batch 1.

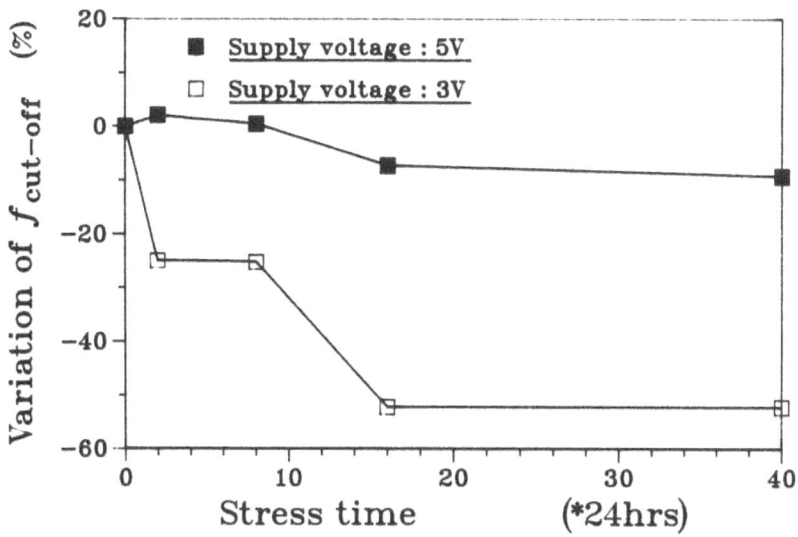

Fig. 4.8. Degradation of the cut-off frequency with voltage–temperature stress at two different supply voltages: 3 and 5 V. The diagram shows a larger effect at 3 V than at 5 V.

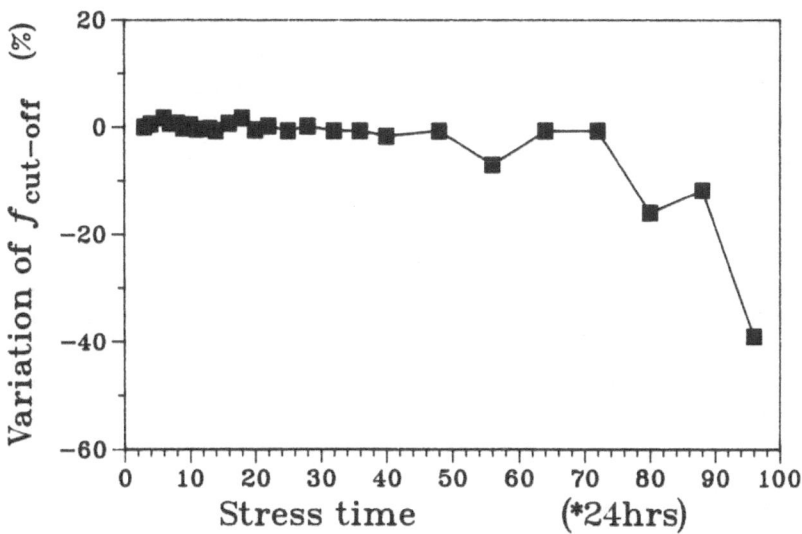

Fig. 4.9. Variation of the cut-off frequency with stress time to illustrate the gradual process of degradation. The device B1 4013 27/1 is here in the toggle mode.

after the first 10 days of stress. This gradual decrease continues for quite a long period of time and then a dramatic decrease starts. This trend has been widely observed in many devices studied. We may conclude that the drastic degradation of the cut-off frequency will occur sooner or later once the trend of the decrease becomes observable.

In the above discussion we did not mention the operation mode of the device being tested. In fact, as we discussed in Section 3.3, the CoFT is carried out in three different operating modes. The example given above is the result obtained in the toggle mode. It is obvious that the cut-off frequency in different operating modes will show different responses to stress since only those defects that reside within the path being tested affect the cut-off frequency in any related operating mode. Any defects not associated with the path being tested are not covered by the test. As an example we plot in Figure 4.10(A) and (B) the variations of the cut-off frequency with stress for a 4013 sample in three operating modes. As is shown in Figure 4.10(A), the cut-off frequency in the set mode degraded far more severely than that of the toggle and reset modes. It is believed that the relatively large decrease appearing in some runs in these two modes are caused by the severe degradation related to the set mode. This interpretation can be extended to a more general case when the two devices in a single chip are considered. Figure 4.10(B) clearly illustrates the different response of the two devices in a chip to stress compared with Figure 4.10(A).

In addition to the general features of the degradation in cut-off frequency with stress, a strange phenomenon has been observed in some samples. That is, after the device starts showing severe degradation, the device may jump between normal and abnormal so far as the cut-off frequency is concerned. An example of this type is B1 4013 29/2. Its characteristics are plotted in Figure 4.11. As we can see in the last few runs, the device is normal in some runs and extremely abnormal in some other runs. However, such jumps did not appear when the device started showing degradation in the beginning of the stress. It seems that such irregular jumps will inevitably bring about total failure of the device.

Besides the irregular jumps between normal and abnormal values, we have also found that the degradation in cut-off frequency may recover if the device is taken out from further exposure to stress and kept in a storage environment. The deep valley appearing in Figure 4.11 originated from about 50 days storage at room temperature. At the end of the storage the cut-off frequency recovered to normal from being severely degraded. However, such recovery is only temporary since after being put under stress again the cut-off frequency starts to degrade immediately.

Instability of the output waveform appears common in devices showing degradation in the CoFT. The detailed noise analysis on the transient supply current has shown close correlation between jitter noise in the transient current and the instability of the output waveform. This point is further discussed in Section 4.3.

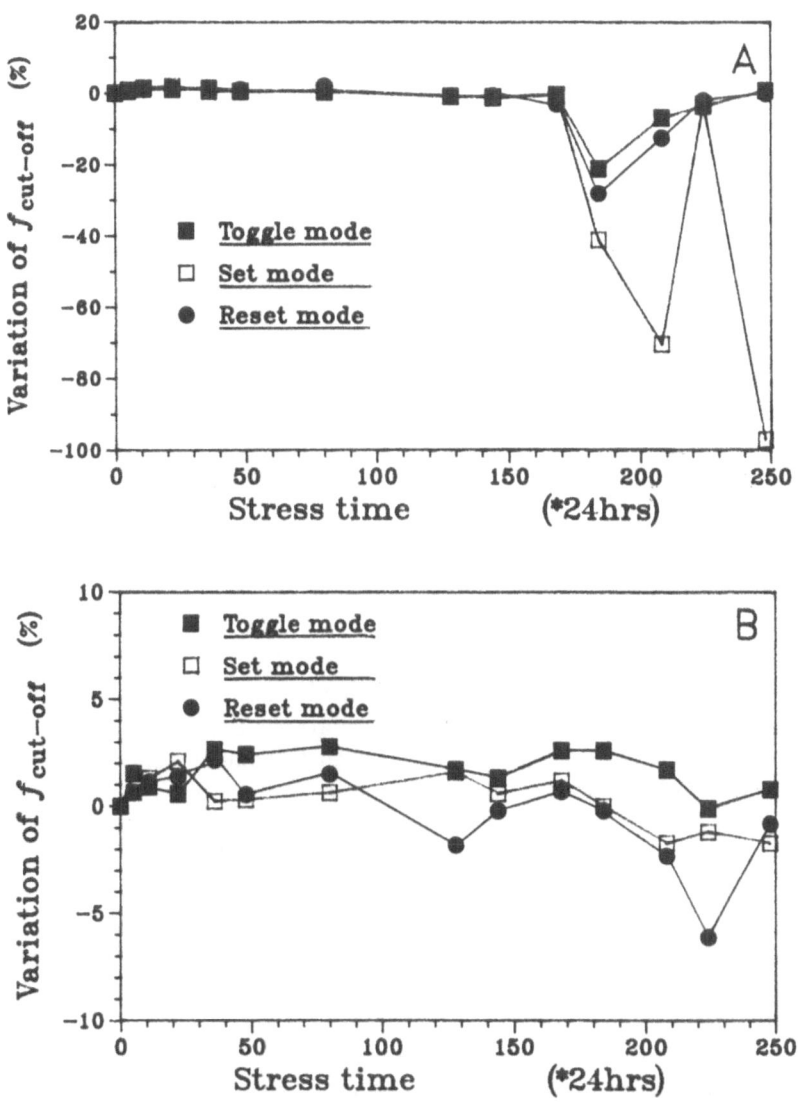

Fig. 4.10. Variation of the cut-off frequency with stress illustrating the response of different operating modes for (A) Device 1 of two devices in a chip and (B) Device 2.

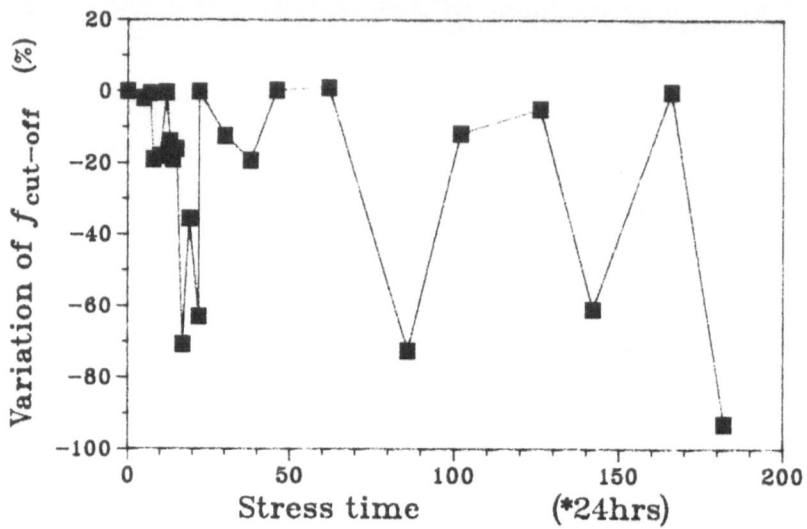

Fig. 4.11. Variation of the cut-off frequency with voltage–temperature stress illustrating irregular jumps between normal and abnormal values.

4.1.4. Discussion of the TrCT

The transient current pulses are only associated with the charging process of the node capacitances and therefore are determined mainly by the channel current of the transistors since the node capacitances are usually of a fixed value which does not change with stress. The charge needed to change a specific capacitance to a predefined voltage level is known and independent of the amplitude of the current. That is, the area under the curve of the transient current pulse will usually be unique. The strength of the current only affects the speed of the charging process. That is, a small charging current will result in a wider and shorter pulse and a large current will result in a narrower and sharper pulse. The TrCT is, therefore, looking for the changes of the pulse width and height rather than area.

It has been widely found that normal devices do not show observable changes in both peak width and height with stress. In general, the variation of the peak width with stress only shows small fluctuations. As an example we plot a typical characteristic of this type in Figure 4.12, where the data were obtained from the second transient current pulse of a normal 4013 operating in the toggle mode. The maximum variation from its mean is about 2 nanoseconds which is the sampling interval.

Besides the general characteristics discussed above, we have observed large degradations in the TrCT on subnormal devices after having been subjected to

108

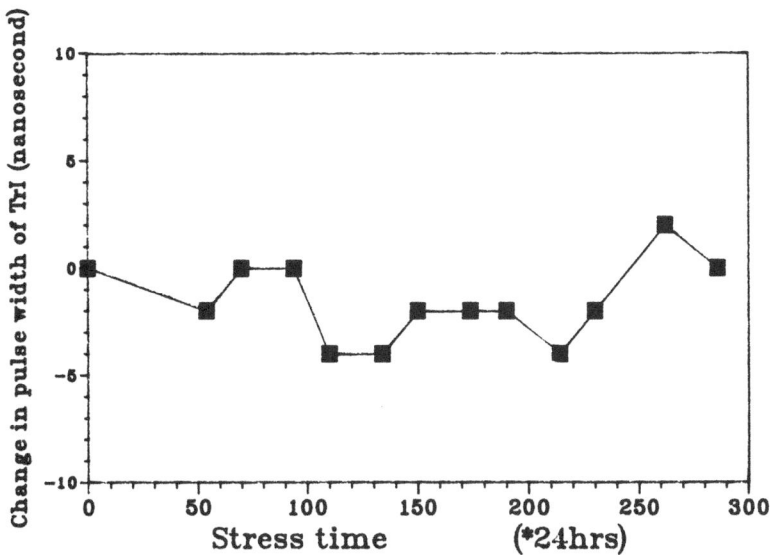

Fig. 4.12. Characteristic variation in width of a specific transient current pulse with stress for a normal specimen.

stress. We first of all discuss some features of the test. Figure 4.13 displays a very general degradation of the transient current pulses. The solid curve is a reference which was obtained in the prestress test and the dotted one was obtained in Run 34/280. It is obvious that the seventh pulse, P7, is degraded with its last peak height decreased and width increased. We also notice that all the other pulses in the diagrams did not change at all and even the first two subpeaks of P7 did not change. Such results are expected since the specific pulse should respond to the switching of a specific transistor pair and therefore a specific defective site should affect only the pulses that are generated by the transistor pair associated with the defect. By referring to Figure 3.27(A) we can see that the seventh pulse is generated when the slave section changes its state so that the output Q is going from low to high and $\bar{Q}$ from high to low. Thus the suspicious defective units are output buffers and their drivers. To provide further evidence we plot, in Figure 4.14, the transient current pulses obtained from the same run and the same device but in the set/reset mode instead of the toggle mode. By referring to Figure 3.27(B) we can conclude that the defect is stimulated when $\bar{Q}$ goes from high to low and is not related to the output Q. By referring further to Figure 3.17 it is not difficult to see that the defect is related to either the first NOR gate of the slave section or the $\bar{Q}$ buffer. By comparing Figures 4.13 and 4.14 it is seen that a difference in the degraded part of the pulses between the two operating modes exists. This leads to the conclusion that the buffer should be free of any 'stuck-at' type defect since otherwise the above mentioned difference

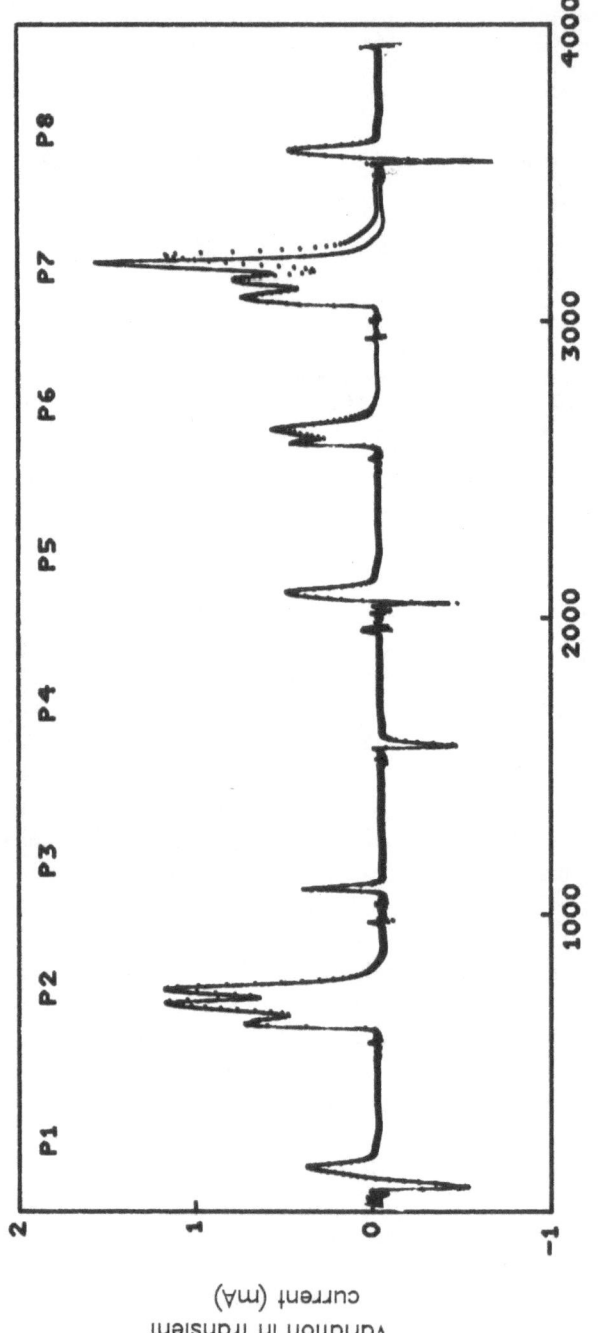

Fig. 4.13. Typical degradation of the transient current pulses after voltage–temperature stress with the specimen working in the toggle mode.

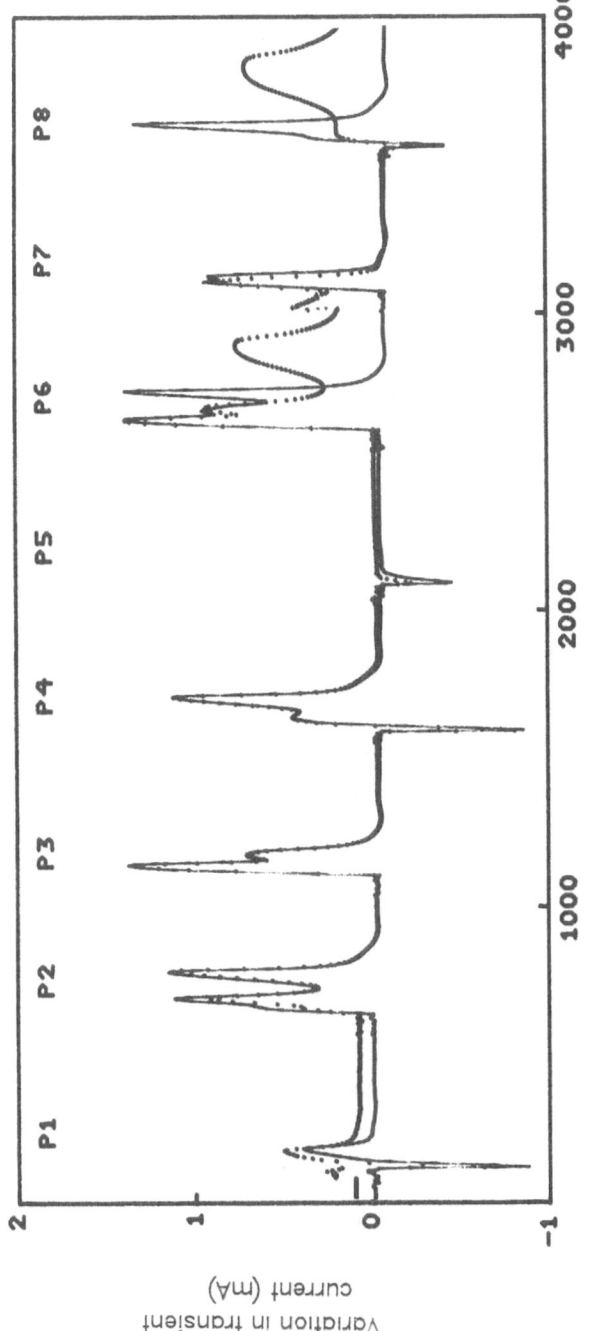

Fig. 4.14. Degradation of the transient current pulses after voltage–temperature stress for the same device shown in Fig. 4.13 but in the set/reset mode.

will not exist. The NOR gate may be responsible for the degradation in the transient current pulse since the different pairs of transistors in the NOR gate work in two operating modes.

In another example we observed similar effects to these described above. Figure 4.15 displays the transient current of a 4013 device in the set/reset modes where the two curves show the reference (solid) and the measured one in Run 37 (dotted). It is apparent that Pulses 2 and 7 are severely degraded. Pulses 6 and 8 are, to some extent, also degraded although it may not be easy to see this from the diagram. Differing from the previous example this device shows a degradation in the transient current when $\bar{Q}$ goes from low to high. The origination of the degradation probably results from a defect that resides in the first NOR gate of the slave section since the first peak of Pulses 2 and 7 which are produced by this NOR gate are not fully developed (referring to Figure 3.17). The slight degradation of Pulses 6 and 8 may be explained as follows. Pulse 6 is produced by the insertion of input SET while the input RESET is held low. The transition first occurs in the second NOR gate of the slave section, its output in turn makes the Q buffer and the first NOR gate of the slave section change stakes and lastly the $\bar{Q}$ is turned over from high to low. The transitions of the first NOR gate involved in Pulses 6 and 7 are produced by two different transistor pairs of the NOR gate. Pulse 8 is produced when the transition of the input RESET occurs from high to low. It is likely that the defect associated with the degradation of this transient current in Pulses 2 and 7 may also affect Pulses 6 and 8. However, it seems that the transient current did not show any changes in the toggle mode although a similar effect as in Pulses 6 and 8 of the set/reset mode is expected.

The above two examples illustrate the degradation of the transient current pulses which occurs only at some peaks of the pulse. However, we have also observed, in some samples, the global degradation of the transient current in all pulses. To illustrate we only plot the first two pulses of the transient current which was obtained from a device in Batch 1 operating in the toggle mode in Figure 4.16. The dotted curve was obtained in Run 42 and the solid one is the prestress reference. It is obvious that both peaks in the pulses decreased in amplitude and increased in width. For clarity we have not plotted all the pulses in the diagram. However all the pulses showed the same effect as did the two pulses illustrated.

A very general feature of the transient current pulses can be observed from the examples given above in that the degradation of the transient current pulses appears as an increase in pulse width and decrease in pulse height. All the subnormal devices showing a degradation in transient current pulses exhibit unique features of this type. This is expected since any degradation in the available channel current to change the node capacitance will result in decreased charging current and therefore in an increase of the charging time.

In both the StCT and COFT the subnormal devices show gradual changes in the transient current with thermal and electrical stress in the beginning of the degradation and drastic changes at a later time. A plot of the general features of

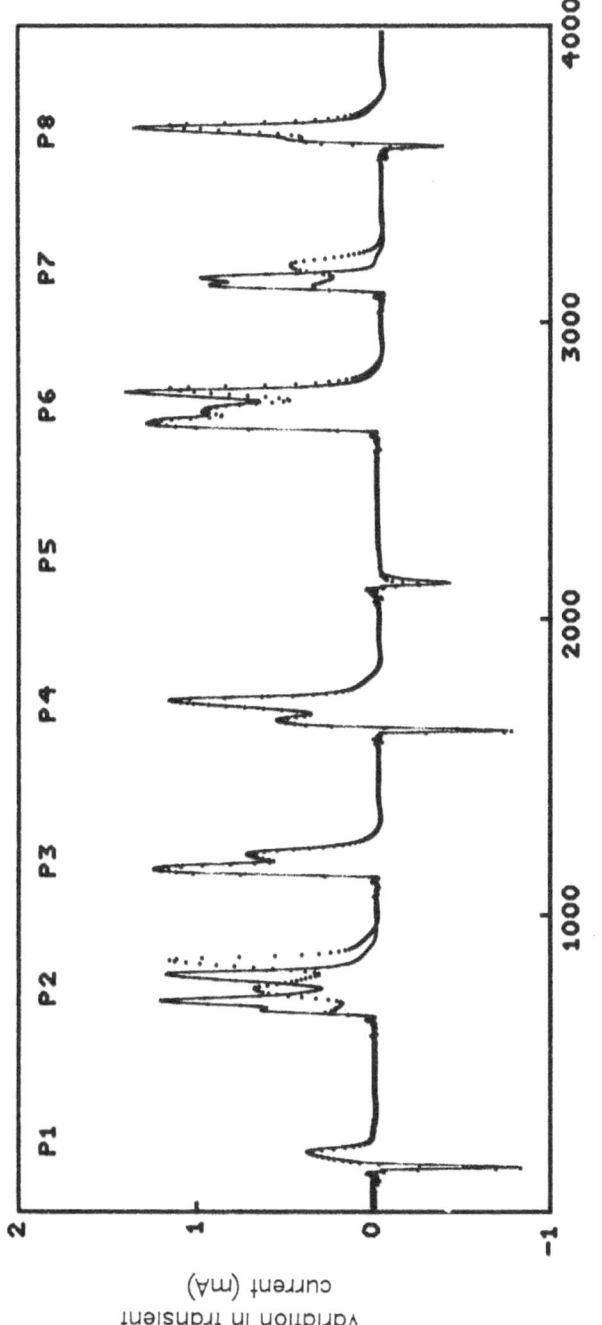

Fig. 4.15. Degradation of the transient current pulses after voltage–temperature stress for a specimen in Batch 1 in the set/reset mode (solid curve—prestress; dotted curve—poststress).

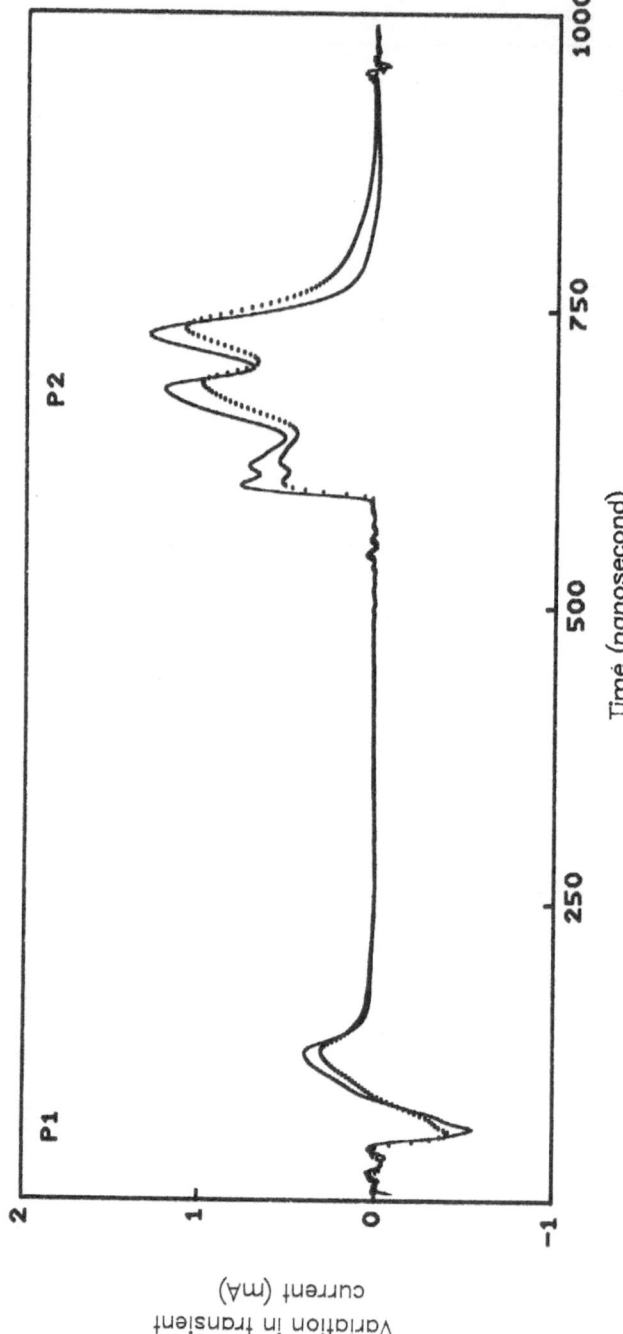

Fig. 4.16. Degradation of the transient current pulses after voltage–temperature stress (solid curve—prestress; dotted curve—poststress).

this type is depicted in Figure 4.17. The data shown in this diagram were obtained from a sample in Batch 1 and in the toggle mode of operation. The solid curve illustrates the gradual changes of the transient current width with stress. The increase at the beginning is only a few nanoseconds. With the progress of the stress, drastic changes of tens of nanoseconds appears. The recovery in the last run seeems to be a very common feature exhibited by the subnormal devices. It has been found that most subnormal devices show irregular changes in the quantities being measured in the tests devised after they started showing large changes in these quantities. However, such a feature is usually not observable in the region where small, gradual changes occur. In addition to the results for Pulse 2 shown by the solid curve, Pulses 3 and 6 are also shown by dotted and dashed curves in the diagram. It is obvious that these two pulses did not show great changes compared with Pulse 2.

4.1.5. Discussion of the StNT

In Section 3.3 we stated that shot noise of the reverse biased p–n junctions is the main noise source when the device works under the static operation condition, and therefore the measured noise in the StNT should be consistent with the shot noise of the measured static supply current in the StCT in both shape of the spectrum and the amplitude of the intensity for a normal device, ie, non-degraded device. Any variation of this noise in the shape of the spectrum or amplitude from that basic level indicated degradation.

It has been widely observed that subnormal devices exhibit excess noise over the intrinsic shot noise after they have been subject to thermal and electrical stress. The $1/f^\alpha$ type of noise with α in a range of 0.5 to 2 appears very common in subnormal devices. A typical spectrum of this type is plotted in Figure 4.18. The vertical axis is the current noise intensity on a logarithmic scale. The exponent α is very near unity at low frequency and the spectrum flattens at high frequency in this example. The horizontal line in the diagram shows the shot noise level of the static current measured. The consistency of the measured noise and expected shot noise at high frequency indicates that the noise sources in this example include both $1/f$ type and white noise components.

It has been found that the excess noise usually appears at some of the test vectors where the defect in the device is stimulated. That is, the occurrence of the excess noise is usually associated with the excess static supply current. To illustrate this we plot the two spectra of the static current noise obtained at two different test vectors in Figures 4.19 and 4.20. The spectrum shown in Figure 4.19 was obtained at test vector 04 of Device 2 and vector 01 of Device 1. It is essentially white and is consistent with the expected shot noise of the measured static supply current. The spectrum shown in Figure 4.20 is obtained at test vector 04 of Device 1 and test vector 20 of Device 1. (Please refer to Table 3.8 for the details of the test vectors.) The $1/f$ type of excess noise appears at test vector 20.

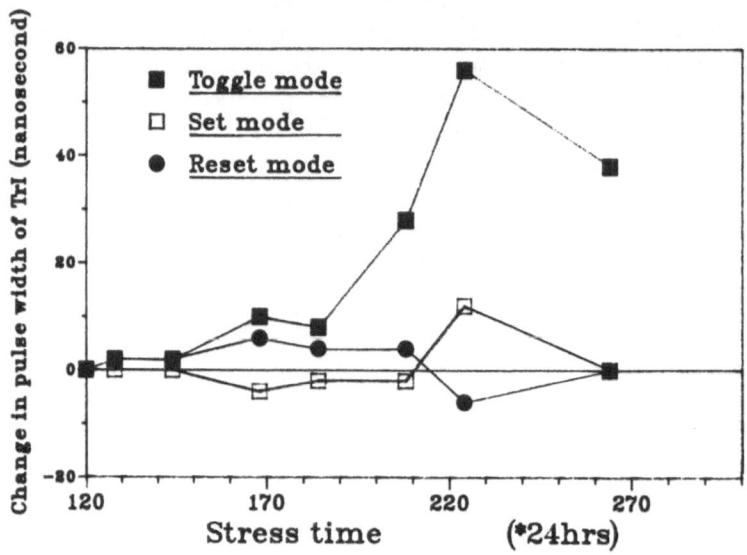

Fig. 4.17.   Degradation of the transient current pulses with voltage–temperature stress for a specimen in Batch 1 and in the toggle mode.

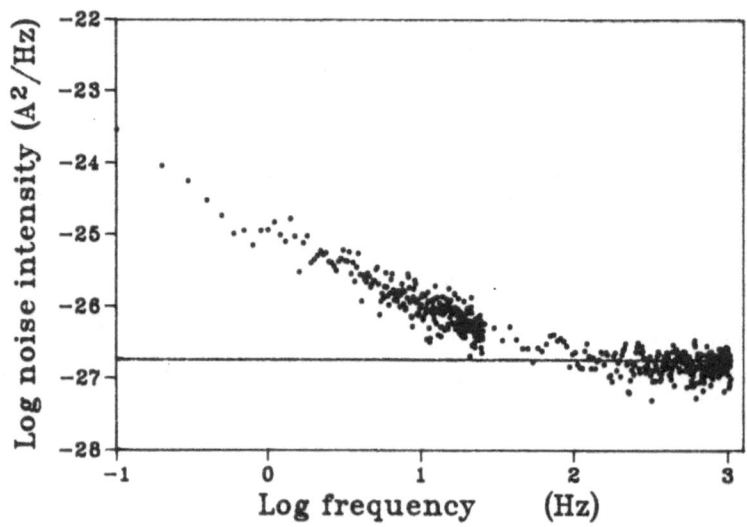

Fig. 4.18.   Typical spectrum of the excess noise in the static supply current for a subnormal specimen, B2 4013 28/1+2. The calculated shot noise level is shown.

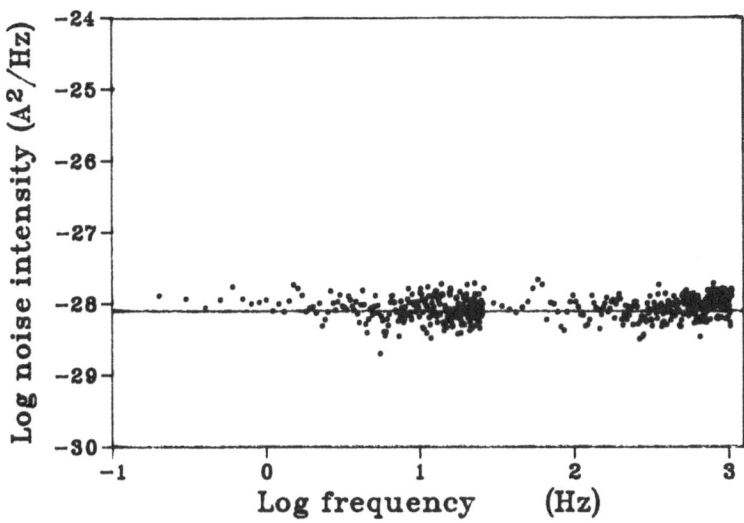

Fig. 4.19.   Spectrum of the current noise for B1 4013 29/1+2 (Run 49) at vector 04 of Device 2 and vector 01 of Device 1.

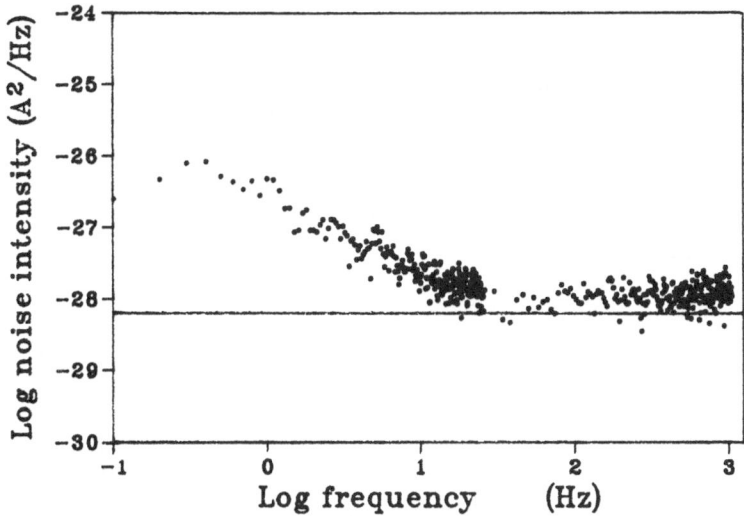

Fig. 4.20.   Spectrum of the excess current noise for B1 4013 29/1+2 (Run 29) at vector 04 of Device 2 and vector 20 of Device 1.

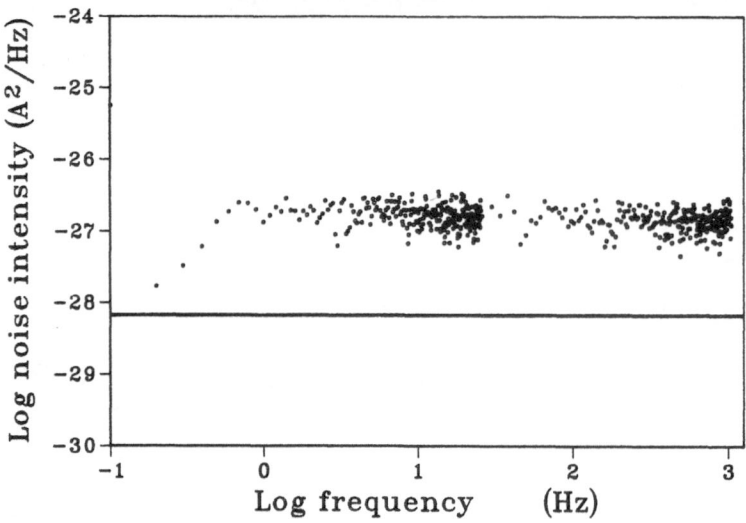

Fig. 4.21. Spectrum of the excess white noise at low frequency for B1 4013 26/1+2 (Run 16).

In addition to the $1/f$ type excess noise observed in most of the defective devices, excess white noise is also observed at low frequency. As shown in Figure 4.21 the noise spectrum is white in the frequency range of 1 to 1000 Hz and the intensity is nearly two orders of magnitude higher than the shot noise. This is possibly a generation–recombination(g–r) noise with a characteristic frequency much higher than the measurement range.

The excess static supply current occuring in the StCT is always accompanied by excess noise occuring in the StNT. This is expected since the defect in the device will always produce excess noise if it does cause excess static supply current of the device. Also, the excess noise is usually more distinctive than the excess current in a specific device. All the examples shown above illustrate this feature.

4.1.6. Discussion of the TrNT

As pointed out in Section 3.3, the TrNT looks at the fluctuation of the transient current with time. Although the amplitude fluctuation of the transient current with time should be observable, we have observed in our work only the fluctuation in timing that causes jitter noise. The general characteristics of the jitter noise in the current pulses are shown in Figure 4.22. The amplitude of the noise intensity is proportional to the slope of the current pulse at the point where the sample gate is set. That is, the variation of the noise intensity with the gating position on current pulse follows the differential of the pulse waveform over time. The maximum value of the noise intensity occurs in the sharp rising and falling parts

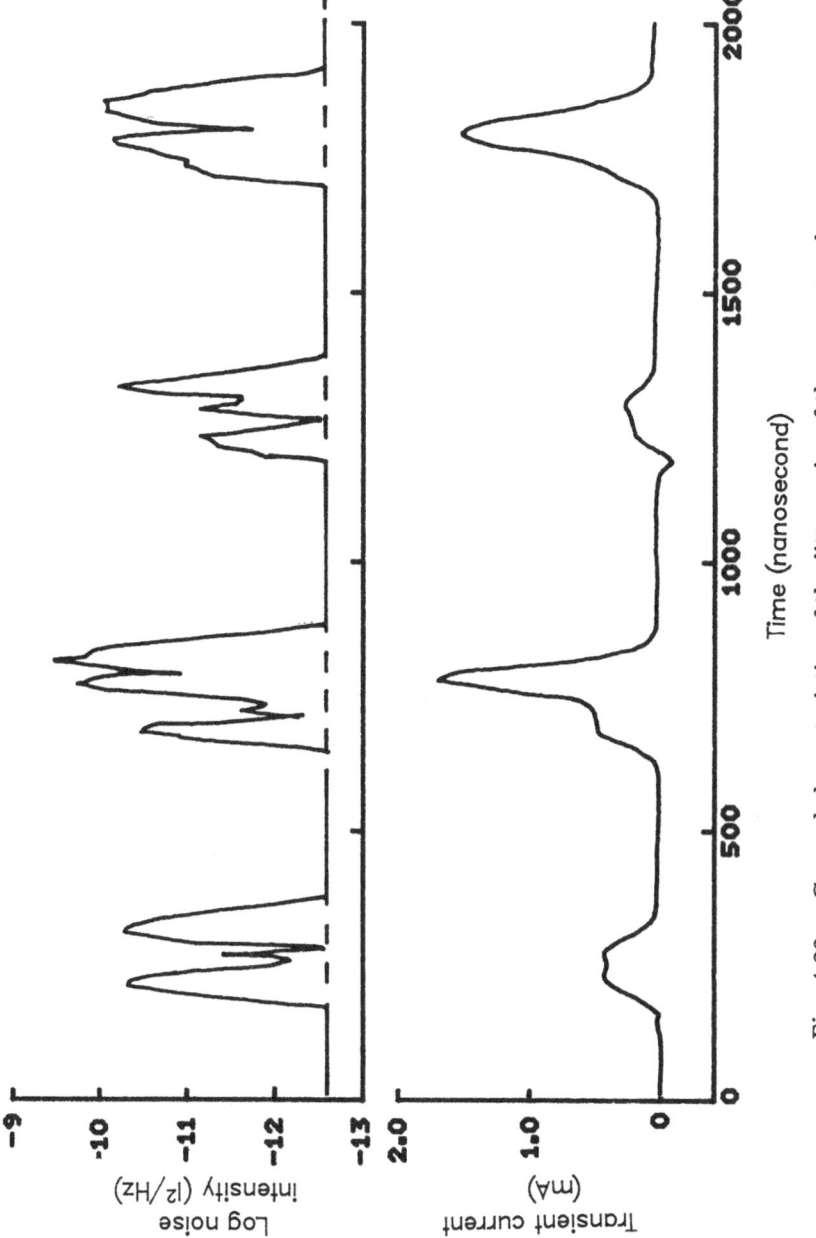

Fig. 4.22. General characteristics of the jitter noise of the current pulses.

119

of the pulse and the minimum value occurs at the points where the peak occurs.

The detailed spectrum of the jitter noise is shown together with the spectrum of the system noise in Figure 4.23(a) and (b). Figure 4.23(a) illustrates the jitter noise at P2 of the current pulses in B1 4013 27/2. This specific spectrum is obtained from the measurements done at the point shown by the cross symbol in the small graph at the top-right side. This is $1/f$ type excess noise with α slightly larger than one. The spectrum shown in Figure 4.23(b) is actually the system noise spectrum since it is the same as the one obtained by sampling the background current where no transient current occurs. Although the spectrum is similar in shape to the one shown in Figure 4.23(a), the noise intensity is much lower than that for the subnormal specimen.

After the analysis of many devices that exhibit jitter noise, it is observed that the noise spectra of this type are very similar regardless of which peak of which device was tested. However, the relative amplitude of the noise intensity is very different from peak to peak, and from device to device.

The occurrence of jitter noise in a subnormal device depends on the test pattern used to produce the transient current pulses. That is, the jitter noise appears only when the test pattern stimulates the defect that produces jitter noise. The following two examples clearly illustrate this feature. Figure 4.24 shows the jitter noise of B1 4013 52/1 when tested in the toggle mode. It is clear that the jitter noise is observable only on the second and fourth peaks. By refering to Section 3.3 we know that pulses one and three are produced mainly by the master section of the flip-flop and pulses two and four by slave sections. Thus the jitter noise occuring in this specific device indicates the degradation of the slave section of the circuit.

Figure 4.25 shows the jitter noise of B1 4013 51/2 in both toggle and set/reset modes. For simplicity, only two peaks of each mode are illustrated in the diagram with P and S indicating toggle and set/reset modes, respectively. The device exhibits the jitter noise only in the toggle mode. This may indicate the degradation associated with the clock on input D.

It is obvious that the jitter noise is caused by an internal device related process rather than external effects, like input signal timing or power supply fluctuations, since otherwise we would not see the different behaviors of each pulse in a device.

The irregular appearance and disappearance of the jitter noise with stress seems common to most of the devices that exhibit jitter noise in the TrNT. A typical example of this feature is listed in Table 4.3. The listed values in a specific run indicate the maximum current noise intensity occurring in the transient current. They are in the order of $\mu A^2/Hz$, and increase gradually with stress. It is almost certain that devices show severe degradation in many of the parameters measured and will fail in a relatively short time if they start exhibiting the above feature.

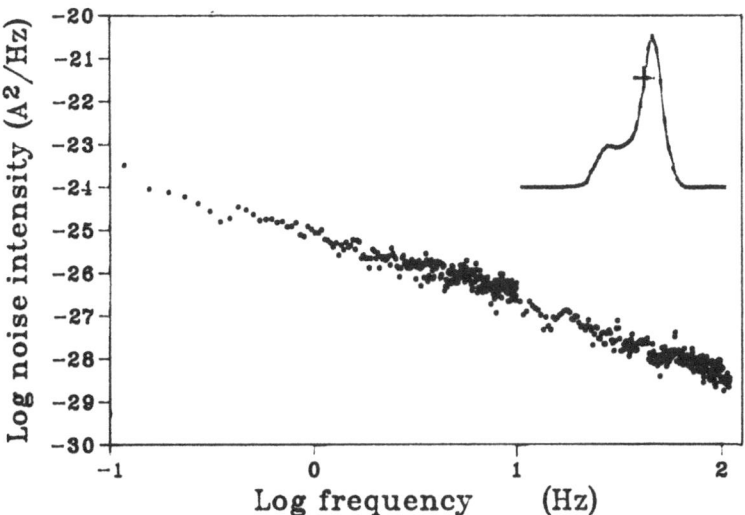

Fig. 4.23(a). Spectrum of the jitter noise on Peak 2 of a subnormal device B1 4013 27/1 which is obtained from the measurement done at the point shown by the cross symbol on the pulse.

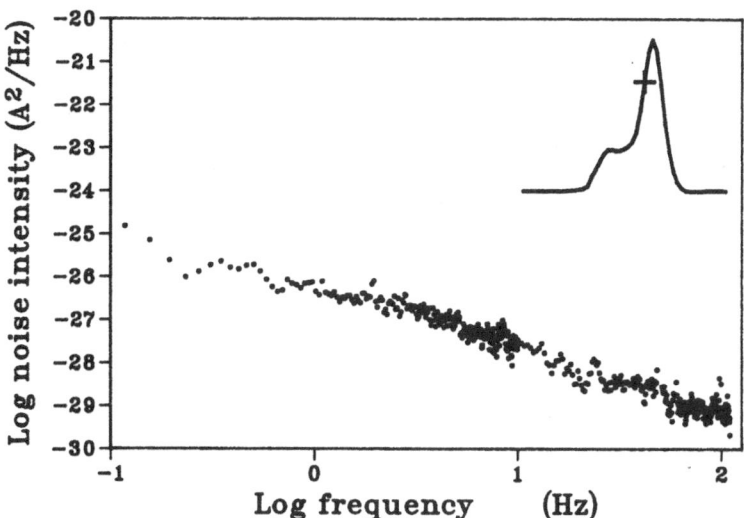

Fig. 4.23(b). Spectrum of the noise on Peak 2 of a normal device B2 4013 27/2 which is obtained in the same way as for B1 4013 27/1 and illustrates the lower value of the system's noise although the spectrum is similar.

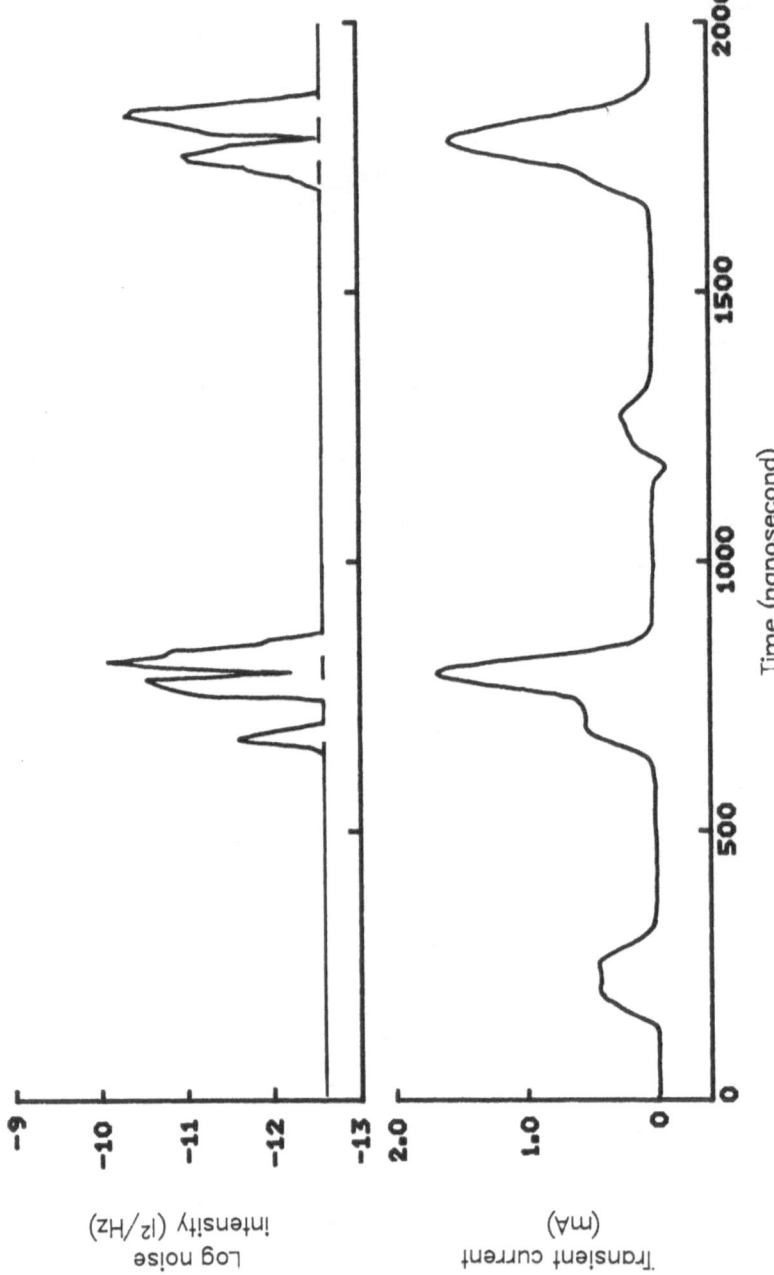

Fig. 4.24. Jitter noise measured in the toggle mode of B1 4013 52/1 which shows that such excess noise only occurs on some specific pulses.

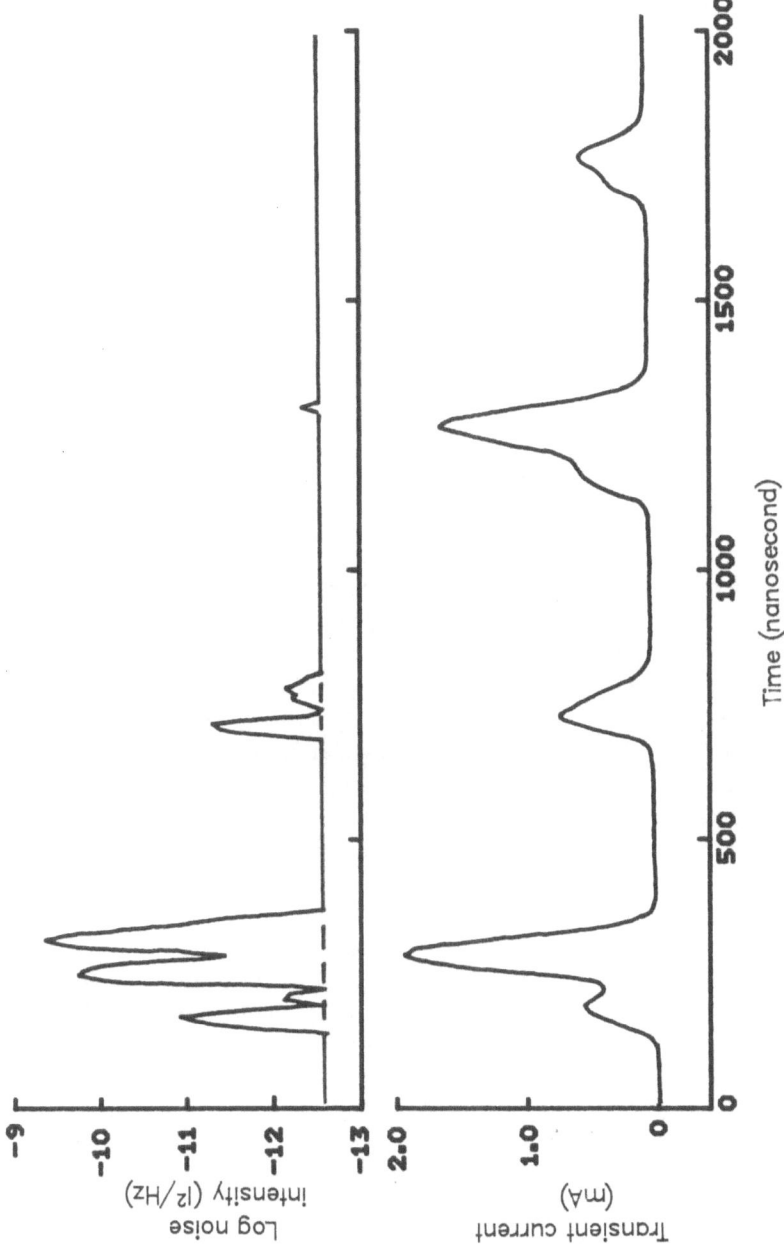

Fig. 4.25. Jitter noise measured in both toggle and set/reset modes of B1 4013 51/2 with 'p' and 's' indicating toggle and set/reset modes, respectively.

Table 4.3. The Variation of the Jitter Noise with Stress for Sample
B1 4013 27/2

Run	Jitter noise (A^2/Hz) at 1 Hz (Max noise intensity)
22/48	4×10^{-12}
23/56	None
24/64	4×10^{-10}
25/72	None
26/80	None
27/88	None
28/96	2×10^{-9}

4.1.7. Discussion of the ThVT

The threshold voltage test reveals the characteristics of the threshold voltages and gain factor of each of the input transistor pairs. Although the number of input transistor pairs is usually very small compared with the total number of transistors in the circuit (especially in VLSIs), the test still gives a very good statistical estimation of the threshold voltage and gain factor of the transistors in the circuit since the input transistor pairs are likely to be distributed uniformly around the edges of the die. The 4014 shift-register is a particularly suitable device to be studied over the whole die. Hence only the 4014s are discussed as examples to illustrate the variation of the threshold voltage and gain factor after they have been subject to thermal and electrical stress.

A good device should not show abnormal changes in both threshold voltage and gain factor with the thermal and electrical stress employed in our work since the electrical stress part of it is dynamic, ie, each node in the circuit will be exercised in both logic-1 and logic-0 states in a similar time duration for both states. This feature has been observed for good specimens in our work. General characteristics of the variation of both parameters with stress in a good specimen are plotted in Figure 4.26, where only the threshold voltages and gain factor of the n-channel transistor in the pair of the serial input, S_{in}, are shown in the diagram. Both threshold voltages and gain factor do not change more than a few percent from the original values. Similar curves will be obtained for other input transistors in the circuit. By doing a statistical analysis on many specimens, it seems that the threshold voltage of the p-channel transistor varies less than that of the n-channel transistor.

Despite the fact that a dynamic stress does not intentionally introduce a shift of the threshold voltage and gain factor of the transistor, the shift of the threshold

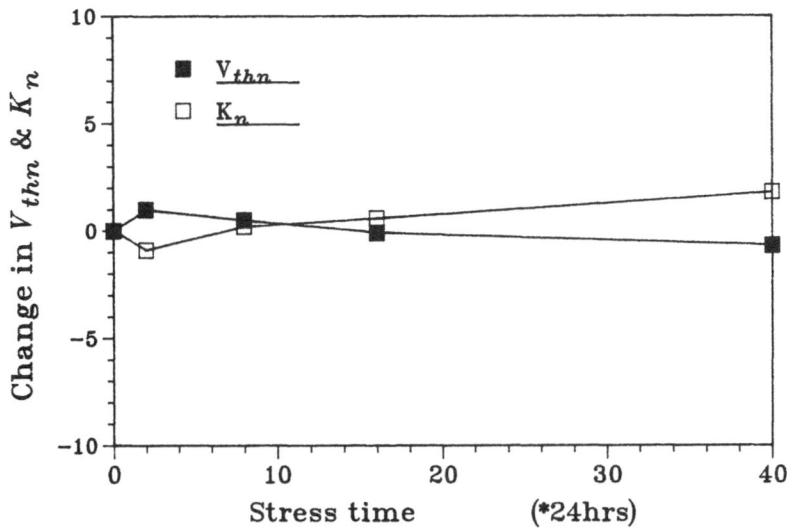

Fig. 4.26. The percentage change of the V_{Th} and K with voltage–temperature stress for a normal specimen in Batch 4.

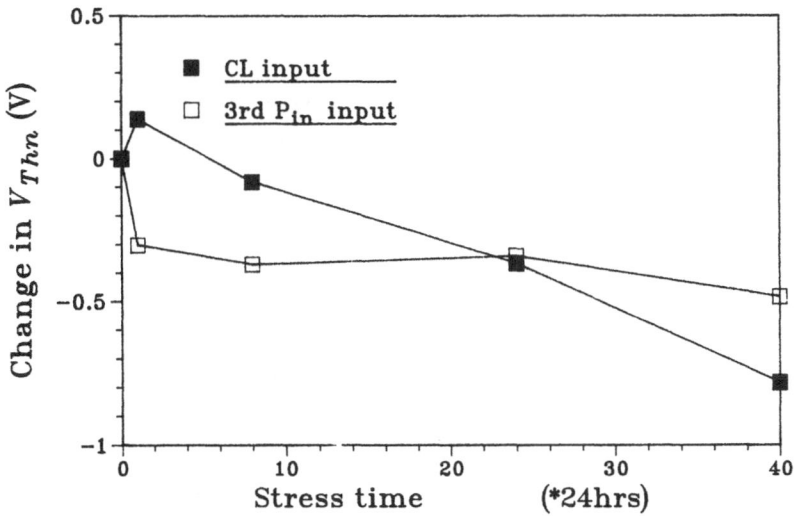

Fig. 4.27. Variation of the V_{Th} for CL and third parallel-in input units with voltage–temperature stress.

voltage and the degradation of the gain factor still occur in subnormal specimens during stress. We present some typical features in the following example. Table 4.4 shows the variation of the V_{Thn} and K_n with stress for all eleven input n-channel transistors in a specimen of Batch 4. Large changes in threshold voltage are observable in some tranistors such as the ones in CL and some of the parallel-in inputs units and such shifts seem gradually to build up with the progress of the stress. We plot the threshold voltages of the n-channel transistors in CL and the third parallel-in units in Figure 4.27. Pins 3, 4, 7 and 8 exhibit very similar trends for the threshold voltage shift with stress as shown in Figure 4.27. This clearly indicates that the abnormal shift of the threshold voltage in the n-channel transistors of these units results from the same mechanism.

Opposite to most input n-channel transistors that show a decrease in threshold voltages, the ones in inputs P/S and the fifth parallel-in inputs show an increase in threshold voltage, although the trend of a further increase faded away in Run 10.

Besides the shift of the threshold voltage, V_{Thn}, the gain factor, K_n, of some transistors also degrades significantly. For example, the gain factor in the serial-in, second parallel-in units degraded up to 50% in Run 10. However, improvement of the gain factor also occurs in some of the units such as the fifth parallel-in unit, for example. Similar to the shift of the threshold voltage with stress, the variation of the gain factor also shows a gradual change as can be seen in Table 4.4.

Despite the irregular changes of the threshold voltage in the n-channel transistors, the p-channel transistors seem very stable in threshold voltage independent of the stress. We have not observed so much change in the threshold voltages in the p-channel transistor as we have observed in the n-channel transistor. Although a negative shift of the threshold voltage in the n-channel transistors has been observed in some of the specimens, the shift is usually only a few percent and is never greater than 15%.

The gain factor of the p-channel transistor shows a similar effect with stress as for the n-channel transistor. It usually degrades, ie, decreases in value, with stress, although the improvement of this parameter may occur in some very rare cases.

Devices like the one discussed above show very different characteristics in the shift of the V_{Thn} and K_n with stress. We have also observed in some of the specimens that the V_{Thn} and K_n values of all the transistors in a specimen may shift in the same direction. Table 4.5 illustrates this characteristic. The shift of V_{Thn} occurs in all the transistors being measured and the amount of the shift is nearly the same for all serial-in and parallel-in inputs. This indicates the unique mechanism that causes the shift of the threshold voltage.

In short, the thermal and dynamic electrical stress usually causes a shift of the threshold voltage in the negative direction, although the shift of the threshold voltage in the p-channel transistor is sometimes less pronounced. Degradation of the gain factor in both n-channel and p-channel transistors is observed, although improvement is also observable in some rare cases.

Table 4.4. The Variation of the Threshold Voltages, V_{Thn}, and the Gain Factor, K_n, with Stress for a Specimen in Batch 4

INPUT	RUN									
	00/0		01/2		05/8		08/24		10/40	
	ΔV_{Thn}	$\Delta K_n(\%)$	ΔV_{Thn}	$\Delta K_n(\%)$	ΔV_{Thn}	$\Delta K_n(\%)$	ΔV_{Thn}	$\Delta K_n(\%)$	ΔV_{Thn}	$\Delta K_n(\%)$
P/S	1.647	4.686	0.222	3.0	0.244	4.2	0.239	3.3	0.233	2.8
CL	1.742	4.643	0.139	0.4	0.082	-0.9	-0.366	-13.8	-0.783	-22.3
S_{in}	1.933	1.117	-0.043	-4.5	0.005	-15.2	-0.124	-36.8	-0.288	-41.8
P_{in} 1	1.941	0.951	0.013	9.3	-0.137	9.8	0.033	12.1	-0.099	3.9
2	1.801	1.972	0.024	-3.1	0.003	-7.2	0.014	-17.6	-0.113	-50.5
3	2.483	0.935	-0.302	8.9	-0.370	12.2	-0.340	15.9	-0.483	7.4
4	2.428	0.965	-0.297	7.1	-0.341	10.1	-0.319	14.6	-0.459	6.4
5	1.643	0.691	0.107	23.3	0.118	29.2	0.271	44.9	0.109	31.2
6	1.858	1.873	-0.007	-4.9	-0.027	-8.1	-0.007	-12.1	-0.108	-24
7	2.524	0.863	-0.242	-0.2	-0.303	1.4	-0.281	5.4	-0.447	-2.7
8	2.508	0.803	-0.274	3.5	-0.329	5.0	-0.297	10.1	-0.478	-0.4

Table 4.5. The Shift of the Threshold Voltage in the Same Direction After the Specimen has been Stressed

I/PS		V_{Thn} (V) Run 00/0	V_{Thn} (V) Run 11/64
P/S		1.606	-0.478
CL		1.621	-0.363
S_{in}		1.834	-0.212
P_{in}	1	1.812	-0.214
	2	1.812	-0.218
	3	1.809	-0.228
	4	1.806	-0.227
	5	1.811	-0.228
	6	1.814	-0.226
	7	1.805	-0.222
	8	1.797	-0.229

4.2. BEHAVIOR OF DEVICES SUBJECT TO IONIZING IRRADIATION STRESS

The purpose of the radiation stress in our work is to provide well controlled variations of the defect parameters which are involved in the tests which have been devised. This is an efficient means of characterizing the electrical properties of the device for reliability use. To this end a high dose rate was employed to speed up the degradation of the device and only a few samples were studied. In the following section we present some general features of the radiation stress in the StCT and ThVT together with the brief discussion of TrCT and COFT on 4014 shift-registers.

A summary of the conditions under which the radiation was conducted is listed in Table 4.6, where two sets of dose rate were employed in the stress.

Table 4.6. Summary of the Radiation Conditions

Set	Dose rate	Supply voltage (V_{dd})	Bias
1	200 Rads/s	10 V	Static operation
2	400 Rads/s	10 V	Static operation

4.2.1. Discussion of the StCT

One of the fundamental effects of the radiation stress on CMOS digital circuits is the increase in the static supply current of the circuit regardless of the details of the stress such as radiation source, dose rate, and so on.

Basically there are two processes associated with the excess static supply current after the devices have been subject to ionizing stress. One is the parasitic effect that usually occurs in the field oxide area. The build up of the interface states and trapped charges in the field oxide will enhance the parasitic surface channels and deteriorate the performance of the reverse biased p–n junction and therefore give rise to excess static current. The other is the subthreshold effect. The channel current in the subthreshold region where the surface potential is in depletion or the weak inversion region varies exponentially with the gate voltage. Hence the static supply current increases dramatically when either of the transistors in the pair is working near the depletion mode, ie, the threshold voltage shifts towards zero. Note that the subthreshold effects are usually not provoked while the shift of the threshold voltage is not comparable to the value of the threshold voltage, whereas the parasitic effects always exist no matter how weak the radiation is. In reality the value of the threshold voltage in a CMOS circuit is normally greater than 1 V and hence the parasitic effect dominates the generation of the excess current unless the device has been subject to a very high dose radiation. Our experiments are consistent with the above discussion. To illustrate this we discuss a specific sample. The values of the static supply current with radiation stress at some specific bias conditions are plotted in Figure 4.28 following a total of 1.58 MRads radiation. Curve a in the diagram is obtained from the measurement done under the same bias condition as used in stress. Curve b is obtained under the bias which is otherwise the same as in the stress except for P/S which is biased at logic-0 (it was biased at logic-1 during stress). Curves c, d, e and f are obtained in a similar way as Curve b with clock, serial-in, first and second parallel-in inputs set to the opposite logic value as in stress, respectively.

Curve a shown in the diagram illustrates the minimum current we can achieve after stress, and this current, to some extent, reveals the gross damage to the circuit such as the damage to the p–n junction between the n-substrate and p-well and the protection p–n junctions that are reverse biased in both measurement and stress since the current measured is the sum of the above component and other current sources. As is shown in the diagram, the value of the excess static current is far below nanoamperes after a total of 1.5 MRads radiation and still within the specification of most manufacturers. Therefore we may conclude that the degradation caused by this type of damage is insignificant so far as the static supply current is concerned.

All the other curves in the diagram, however, exhibit very different characteristics as compared with Curve a. This is expected since the bias conditions under which the device is stressed and its static supply current is measured are totally different between Curve a and all other curves. These curves show a rapid increase of the current with stress. Note that this dramatic increase of the static current results solely from the specific unit containing the complementary transistor pair that is biased to the opposite logic value to that during the stress since all other transistor pairs are biased in the same way as in the stress.

We may have noticed that Curves b and c, e and f are very close to each other. This is always true and the relative positions of all the curves are always the same for all devices studied. This is not surprising if we take the device structure into account. The nominal gate widths of relative input transistor pairs are listed in Table 4.7. It is clear that the excess static supply current produced is proportional to the sizes of the devices that generate the excess current by comparing the curves shown in Figure 4.28 and the data shown in the list.

Table 4.7. The Nominal Gate Widths of the Transistors in the Input Unit

Input	Nominal gate widths (nm) (p/n)
P/S	64/64
CL	64/64
S_{in}	20/20
P_{ins} 1 − 8	40/30

Taking a closer look at the two pairs of curves in the diagram we may notice that Curves e and f are closer than Curves b and c, although the structural dimensions are the same for P/S and clock. The deviation between Curves b and c results very possibly from the bias condition under which the device has been stressed or tested. The input P/S is biased at logic-1 and the clock is biased at logic-0 while stressed. The current measurements are done under the biases that are opposite to that of in stress. This indicates that the excess static supply current of an inverter is larger when its input is biased at logic-1 than that when biased at logic-0. Note that this result should not be confused with the concept of the worst-case radiation since the mechanisms associated with these two concepts are totally different.

In the concept of the worst-case radiation, the functional performance is considered. That is, the shift of the threshold voltages plays the main role there. In the context we discussed above the transistors are still working in enhancement mode, although a shift of the threshold voltages occurs. Thus, only the parasitic effects dominate the process rather than the subthreshold effect.

Before we end the discussion of the StCT we give a brief account of the $I \sim V$ characteristics observed in the devices that have been subject to the radiation

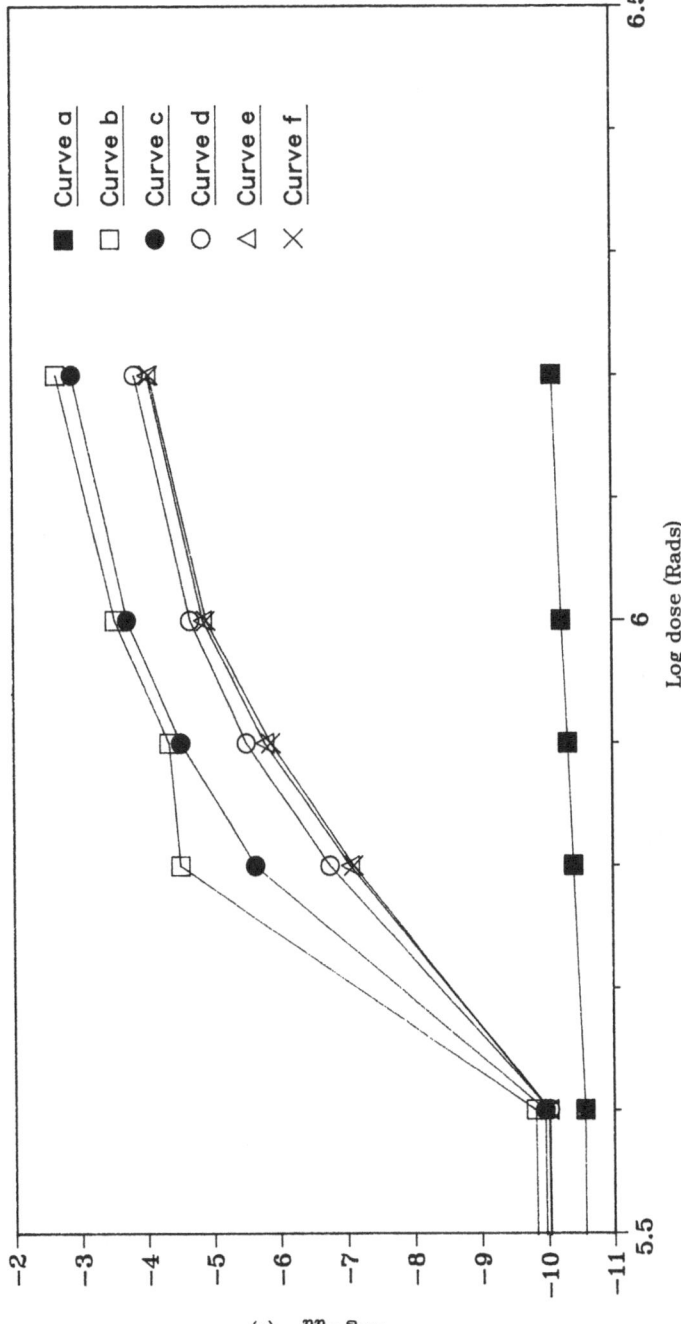

Fig. 4.28. Variation of the static supply current with radiation stress under different bias condition (see text).

stress. It has been found that the turn-on of the excess static current occurs below 3 V and the excess static supply current does not change much with an increase of the supply voltages above 3 V. Figure 4.29 illustrates a typical $I \sim V$ characteristic of this type. The data were obtained in test vector 01 of Run 04 for Device B3 4014 09.

4.2.2. Discussion of the ThVT

The shift of the threshold voltage in MOS transistors results from the build up of the interface states at the silicon dioxide–silicon interface and the accumulation of trapped change in the silicon dioxide. The latter is the dominant factor of the shift in threshold voltage at high dose rate of radiation.

The general characteristics of the shift in threshold voltage with radiation stress are plotted in Figure 4.30. The data in the diagram were obtained by monitoring a specific input inverter in a 4014 circuit that was subject to successive stresses.

As is shown in the diagram, the n-channel transistor shows a decrease in threshold voltage with stress and the p-channel transistor shows an increase in the absolute value of the threshold voltage. As we mentioned earlier, the oxide trapped charges are the main contributors to the shift of threshold voltages in high dose rate stress and they are usually positive changes. Thus, the build up of such charges in the gate oxide area will cause the threshold voltages in the n-channel transistor to decrease and in the p-channel to increase in absolute value.

As we may notice in the diagram, the shift in the n-channel transistor is far more than that in the p-channel transistor. This occurs because of the bias condition under which the device is being stressed. The results shown were obtained from the input P/S inverter with its input biased at logic-1 while being stressed. That is, the n-channel transistor is biased ON and the p-channel transistor biased OFF. Thus, a large electric field is built up across the gate oxide of the n-channel transistor. This electric field enhances the ionization process of the radiation and therefore the build up of the oxide trapped changes. Also, the electric field forces the accumulation of the positive oxide charges near the silicon–oxide interface in the oxide. It again enhances the effect of the oxide charges on the shift of the threshold voltage in the n-channel transistor. However, no electric field is built up across the gate oxide of the p-channel transistor and therefore there are no electric field enhanced effects occurring. Hence, we see a large shift in the n-channel transistor but not in the p-channel transistor.

Because of the effect of the oxide electric field, it is obvious that the characteristics of the shift in threshold voltage will be different if the monitored input inverter is biased at logic-0 instead of logic-1 as discussed above. As far as the n-channel transistor is concerned, it is similar to the p-channel transistor in the above case. There is no electric field across the gate oxide. As we will see in the following examples, the shift of the threshold voltage in the n-

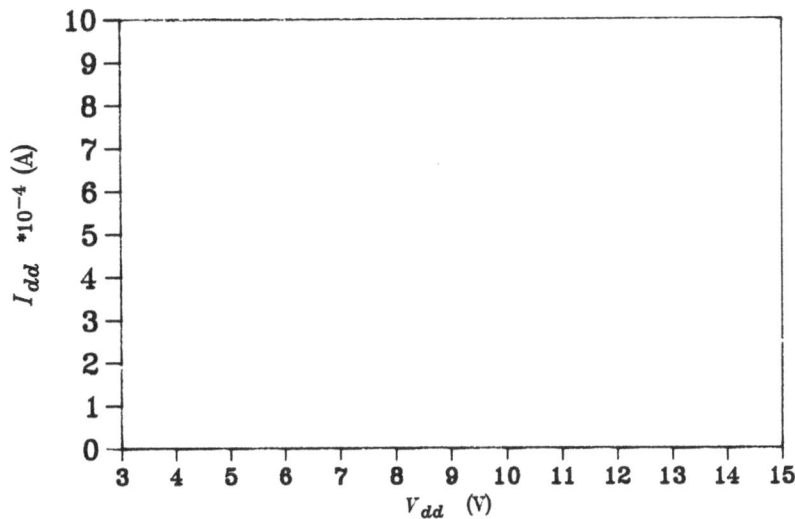

Fig. 4.29. Typical $I_{dd} \sim V_{dd}$ characteristics after the specimen has been exposed to radiation stress.

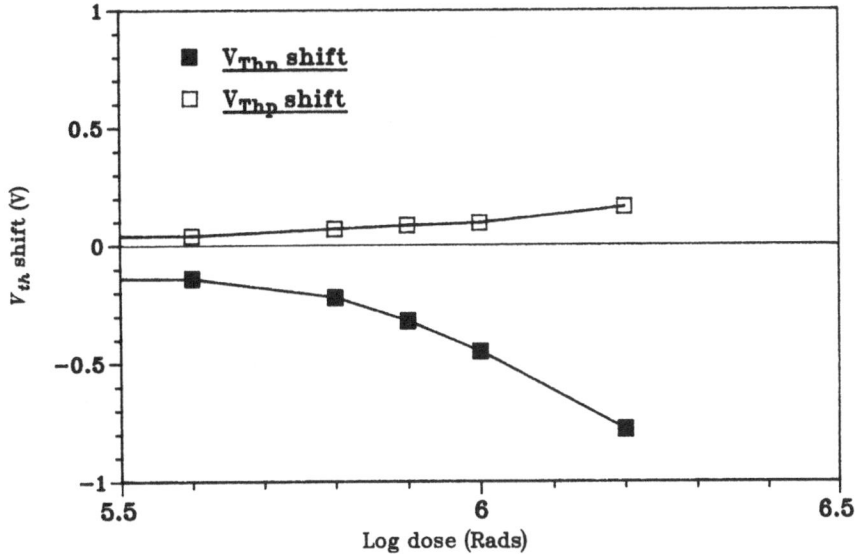

Fig. 4.30. General characteristics of the shift in V_{Th} with radiation stress.

channel transistor is really less pronounced than in the above case. However, the situation for the p-channel transistor is somewhat more complicated. The existence of the electric field across the gate oxide of the p-channel transistor enhances, on one hand, the process of ionization produced by radiation, but drags, on the other hand, the trapped oxide changes away from the silicon–oxide interface. Thus the two processes cancel each other as far as the shift of the threshold voltage is concerned, although the net effect should normally favor the increase of the threshold voltage in absolute value. The experimental evidence shows that there is not a unique characteristic for the shift in threshold voltage of the p-channel transistor as of the n-channel transistor. The typical characteristics of the shift in the threshold voltage with stress are plotted in Figure 4.31 for the input inverters that were biased at logic-0 at its input while being exposed to stress. The characteristics of the n-channel transistor, as we predicted before, are similar to these shown in Figure 4.30 with the absolute shift being less pronounced. However, the characteristic of the p-channel transistor is very different from the one shown in Figure 4.30. It shows an initial increase and then a gradual decrease. How this should be explained is open to question. The possible processes associated with this outcome may be as follows. In the beginning of the stress, the production of the oxide change is the dominant process over other ones so we see an increase at the beginning of the stress. Then, the other two processes may become competitive with the above process. One is the gradual build up of the interface states at the silicon–oxide interface and the other is the gradual accumulation of oxide charges in the oxide near the gate–oxide interface caused by the external electric field. If this is the case we then should have the threshold voltage decrease as is shown in the diagram.

To avoid the effects discussed above being dominant, we have deliberately increased the dose rate from 200 to 400 Rads/second to do the same test as above. The results are shown in Figure 4.32. Because of the increase of the dose rate, the production of the oxide changes are enhanced so the decreases occurring in Figure 4.31 disappeared. This supports, to some extent, the discussion above. In addition, the shift in the threshold voltage of the p-channel transistor is far more than that of the n-channel transistor in this case. By getting these results we are able to confirm that the external electric field applied to the gate oxide enhances the process of ionization caused by radiation. The accumulation of the oxide charge at the other end of the oxide near the gate caused by the external electric field weakens the effect of the oxide charge on the shift of the threshold voltage in the p-channel transistors. The build up of the interface states may also play a role in the shift of the threshold voltages when a relatively low dose is applied.

Besides the test of the threshold voltage, the gain factors of the transistors were also studied. It has been found that the gain factor usually does not vary much with radiation stress in a relative short period of time. The maximum decrease of the gain factor observed after a total of 1 MRads dose radiation, ie, a few hours exposure to stress, is less than 10%. This finding is consistent with

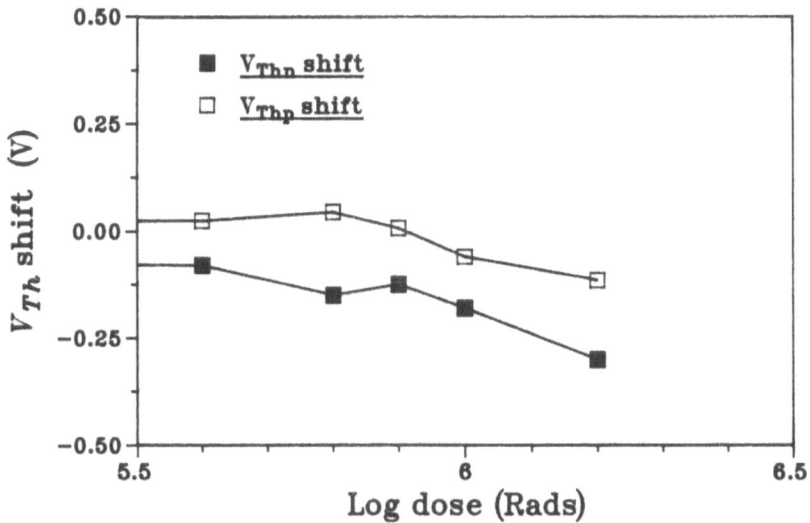

Fig. 4.31. Typical shift of V_{Th} with radiation stress for an inverter that was biased at logic–0 while being exposed to stress.

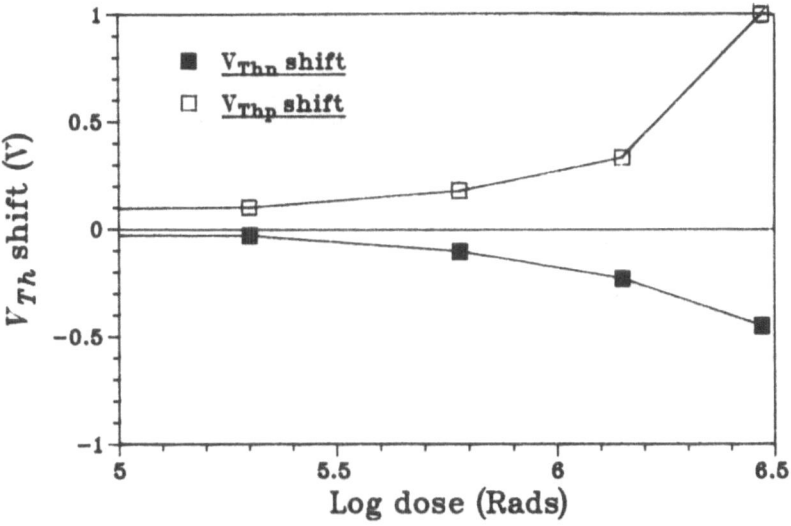

Fig. 4.32. The same plot as in Figure 4.30 but with a dose rate of 400 instead of 200 Rads/second.

the discussion held above where we believe the build up of the interface states in high dose rate but short time exposure stress is signficant so that the degradation of the carrier mobilities in the conducting channel is insignificant and therefore the gain factor does not degrade much.

4.2.3. Discussion of the TrCT

As we pointed out in the previous section, the pulse width is the key parameter to be studied in the TrCT since it is very sensitive to the changes in the available channel current that charges the node capacitance.

It has been found that the variation of the transient current pulses with stress shows quite different characteristics on devices that have been exposed to the two stress environments that were summarized in Table 4.6.

The devices that have been exposed to the stress environment of Set 1 in Table 4.6 show continuous decreases of the pulse width with stress, although the trends fade away gradually at high dose, and some increase of the pulse height at low dose which then decreases with further stress. To illustrate this feature we plot in Figure 4.33 the transient current waveforms of both prestress and poststress tests, where the poststress data were recorded after a total of 0.79 MRads radiation and represented by the dotted curve in the diagram. The difference between the prestress and poststress on the transient current waveforms may not be easily seen from the diagram but it is certain that the width of the pulse is decreased and the height is increased. This result is consistent with the discussion of the previous test. The radiation causes a decrease of the threshold voltage of n-channel transistors and a small change of the threshold voltage of p-channel transistors as is shown in Figures 4.30 and 4.31. It is obvious that the decrease in the n-channel transistor will be most pronounced in this case. This explains why we get the results shown in Figure 4.28 since the decrease of the threshold voltages in the n-channel transistors will give rise to an increase of the channel current in the n-channel transistor.

This also explains the effect of the threshold voltage shift in the p-channel transistor. It is obvious that the increase of the threshold voltage of the p-channel transistor in absolute value appearing in Figure 4.25 shows an insignificant effect on the transient current compared with the shift of the n-channel transistor.

However, the devices that have been exposed to the stress of Set 2 listed in Table 4.6 exhibit very different characteristics for the variation in transient current with stress. A typical result of the degradation in transient current after a total of 1.04 MRads radiation under the dose rate of Set 2 in Table 4.6 is plotted in Figure 4.34, where the dotted curve is from poststress and the solid curve from prestress tests. The variation of the transient current in this case is totally different from that of Figure 4.33. The width of the pulses is greatly increased and the height is decreased which is opposite to the situation occuring in Figure 4.33. Such a difference is believed to be because of the dose rates used in two sets of stress. The increase of the threshold voltages in absolute value in

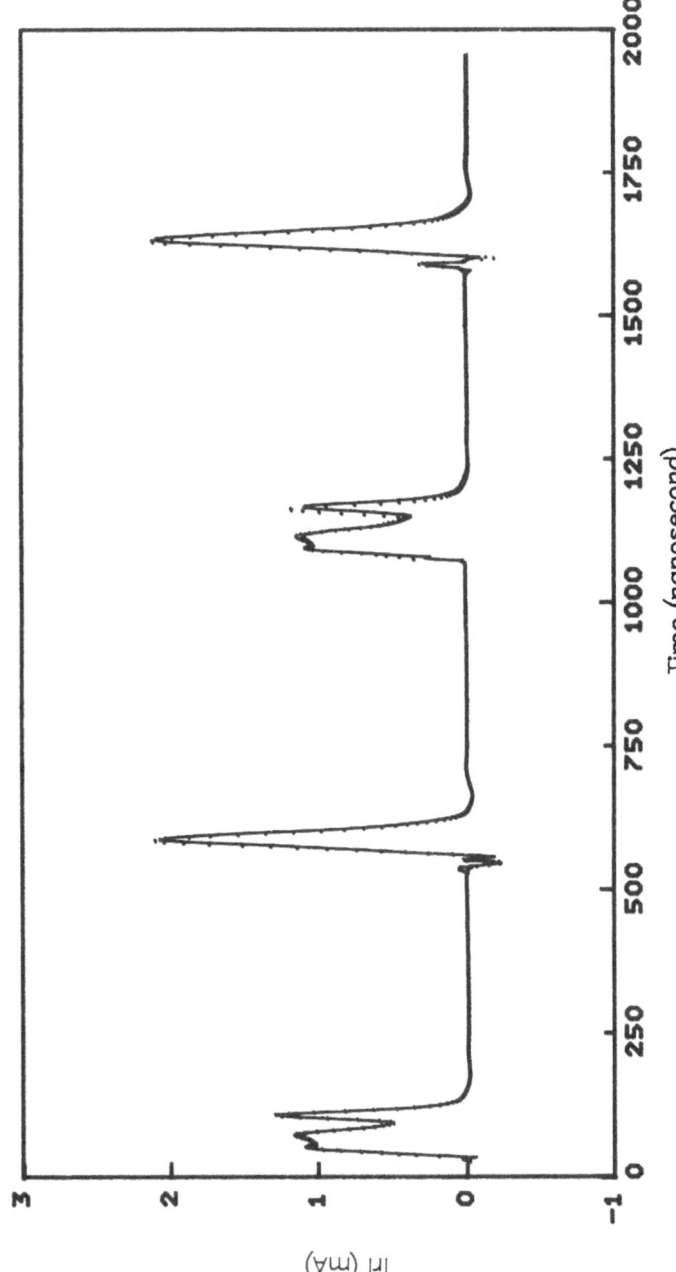

Fig. 4.33. Transient current waveforms of both prestress (solid) and poststress (dotted) tests, where the poststress data were obtained after a total of 0.79 MRads radiation with a dose rate of 400 Rads/second.

137

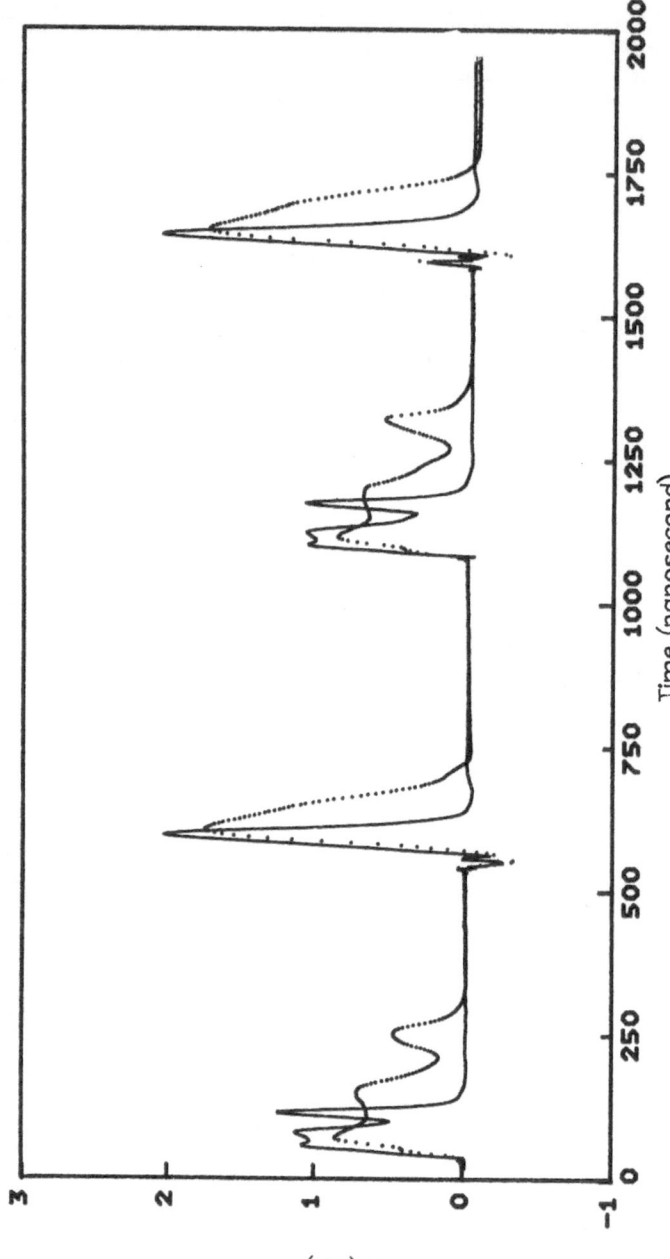

Fig. 4.34. Same as in Fig. 4.33 but with a dose rate of 200 instead of 400 Rads/second.

the p-channel transistor appears quite significant compared with the n-channel transistor when the high dose rate is used in the stress. Such increase inevitably degrades the channel current of the p-channel transistor and therefore causes the degradation in Figure 4.34 to appear.

In addition, we may notice in Figure 4.34 that the area under the transient current waveform is different. This seems contradictory to the assumption that this area should not usually change. However it is not, since that assumption was made under the condition that the channel current is fully applied to change the node capacitance. After the stress the mismatching of the transition point between sucessive stages becomes serious. The difference in the threshold voltages of one type transistor in two successive stages can be as much as 100%. This mismatching will result in a current flowing in the path where both n-channel and p-channel transistor in a pair are turned on, which does not occur in a normal device or a defective device without such mismatching.

4.2.4. Discussion of the CoFT

The CoFT looks at the overall performance of the device working at high frequency. It should, in principle, be similar to the delay time test that has been proposed by some people, but it seems that (from experimental evidence) they are not identical in some respect. In this subsection we mainly discuss the general features of this test and comment on the relationship between our test and the delay time test.

The general response in the cut-off frequency of the device with radiation stress is plotted in Figure 4.35. The cut-off frequency is measured at two supply voltages which are the minimum and maximum supply voltages specified for 4000 series devices. The cut-off frequency at 3 V shows an increase in the beginning of the stress and starts to decrease with further stress. Before failure in functionality occurs, the device degrades severely in the cut-off frequency. It has been found that all the specimens that have been exposed to radiation stress show very similar trends, with an initial increase, a gradual change in the middle as well as a sharp decrease before the total failure of the device.

However, the response of the cut-off frequency at a supply voltage of 15 V is very different from that at 3 V. As is also shown in Figure 4.30, the cut-off frequency shows a trend of gradual increase, although this trend may not be as obvious from the diagram. It showed no sign of any decrease after the device has been exposed to a total of 1.58 MRads stress, which is totally different from the trend observed at 3 V.

Before we give the detailed explanation on the above observation we first discuss the data obtained in the delay time test. Figure 4.36 illustrates the variation of the delay time with stress for the same specimen discussed above. By comparing the two diagrams shown in Figures 4.30 and 4.31 the following features can be observed.

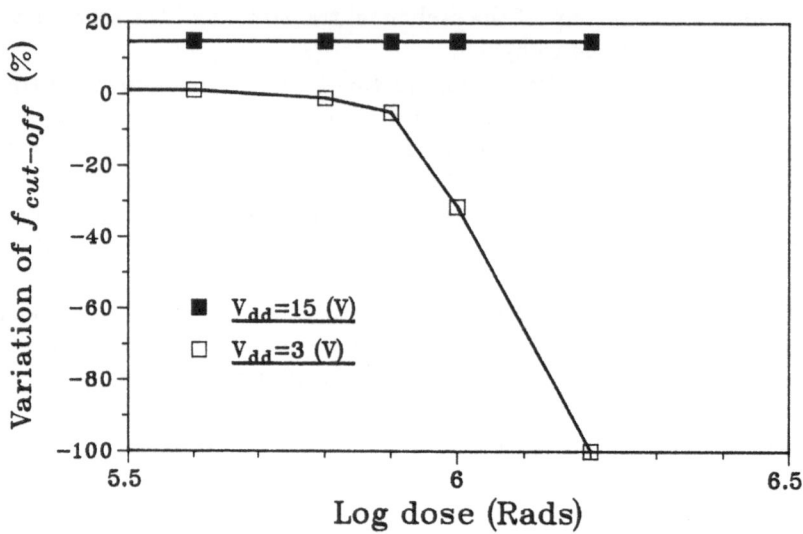

Fig. 4.35.   Degradation of the cut-off frequency with radiation stress for a specimen in Batch 4.

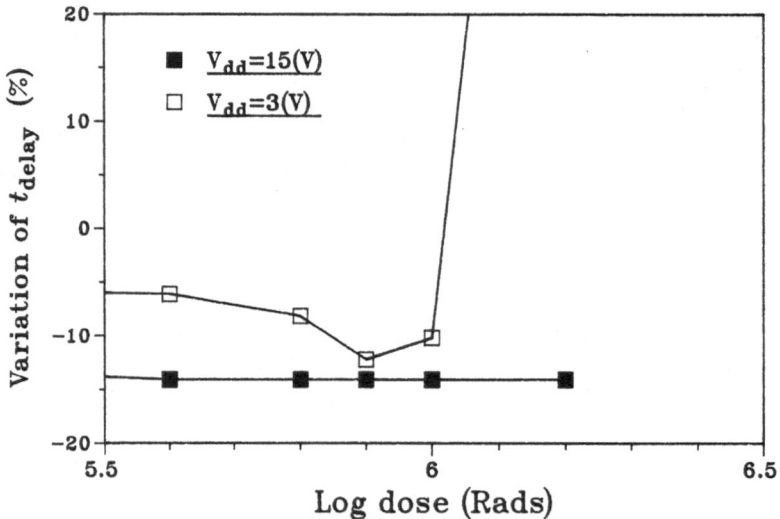

Fig. 4.36.   Degradation of the delay time with radiation stress.

Both the cut-off frequency and delay time show almost straight lines and do not vary with stress. This is expected since the main effect of the radiation stress on the device is the shift of the threshold voltage in a scale of volts which is usually far smaller than 15 V and therefore the $(V_{gs} - V_{Th})$ is almost independent of V_{Th}.

As regards the difference occurring between the prestress and poststress test in both cut-off frequency and delay time at 15 V supply voltage, it seems quite peculiar since such difference is not expected according to the above discussion and has not actually been observed in other specimens.

The cut-off frequency at 3 V supply voltage shows a very different characteristic compared with that at 15 V as shown in Figure 4.30. The initial increase appearing at low dose can be explained as follows. After being exposed to a low dose stress, the device shows a significant negative shift in only the threshold voltage of the n-channel transistor, but not in that of the p-channel transistor. The shift, in turn, results in the improvement of the channel current in the n-channel transistor since $(V_{gs} - V_{Th})$ is increased. This result is consistent with the small decrease in delay time after the device has been exposed to the initial stress. With further exposure to stress the device seems to show contradictory characteristics in delay time and cut-off frequency. The device shows further decrease in delay time and also decrease in cut-off frequency after the initial increase, despite the fact that the decrease in delay time should correspond to the increase in the cut-off frequency. It is surprising that this inconsistency has been observed in all specimens studied. How it is interpreted is an open question. A further discussion on this is given in the next subsection.

In addition, the delay time starts to increase at very high dose as is shown in Figure 4.31. This appears because of the mismatch of the threshold voltages between the successive stages that was discussed in the previous subsection. As shown in Figure 4.34, the excess current flows through the n-channel and p-channel transistors in a pair which, in turn, causes the degradation in both height and width of the transient current pulses. Such degradation affects the delay time.

4.3. ASSESSMENT OF THE TESTS

The previous sections have described the tests, what defects they can detect, how these tests are performed and the typical results obtained for normal and subnormal devices. In this section we discuss whether the tests are valuable in practice and how they compare with each other. It must be remembered that the experiments have been performed in a specific way which may not be totally typical of the requirements of a particular industry. That is, normal production devices have been taken and subjected to a temperature–voltage stress over a period of time. The tests have been performed at intervals during that stress period and the measure of the test has been used to determine whether the device has a defect. The assumption in this procedure is that the defects developed during this experiment are typical of the defects occurring in a device soon after manufacture.

4.3.1. Detection of Defects

A very large amount of information has been obtained in these analog tests on many devices over a long stress period. Here only a summary of the information can be given. Many of the devices showed no change in performance on any test over the whole period of the stress. Most samples still performed within specification after the full stress time, although their performance had deteriorated and abnormal behavior had been detected very early in the stress. This indicates that the tests are extremely sensitive indicators of deterioration.

The test measurement procedures have developed over the time of the project so that complete records of each test on each specimen are not available. Also, the test procedure only called for sampling of specimens within a batch until an abnormal test result was detected in a particular device. After that time a full test was carried out.

Examples of the test history of particular devices are shown in Figure 4.37. These are samples of Batch 1 which have given indication of failure but most still perform to specification at room temperature. To simplify the results a three-level display is given of the analog values obtained during the tests. These correspond to a normal indication equal to the prestress test value, an observable change which is significantly larger than the fluctuation of the test value obtained for a normal specimen and an abnormal change at some arbitrary threshold for each test. Where no test was performed a dotted line is drawn.

The 4013 devices are dual flip-flop so that some tests measure some aspect of the properties of both devices on the same chip. These tests are shown in the centre of each figure while the data specific to each half are shown on either side. The compiled data are shown as StCT and ScNT. These are simplified tests from those described earlier and are the tests with each device in the same logic state, just the test vector 1. This is a very crude test since any defect which occurs may, or may not, be excited by this particular test vector. Only by proceeding through all test vectors for both devices can full information be obtained.

It can be seen that in all these examples some of the tests gave very early indication of failure and other tests soon also gave indications. The observable change sometimes gave a long forewarning but also in other cases the onset of the change was rapid and an abnormal change developed almost at once.

Another common feature of these diagrams is that the results of the tests changed in both directions between the three levels. In particular, abnormal specimens returned to normality. This is a fair assessment but is perhaps overemphasized here since often a device which showed some form of abnormality would be removed from the test procedure and left at room temperature for several days. During this time it might recover its former properties. The

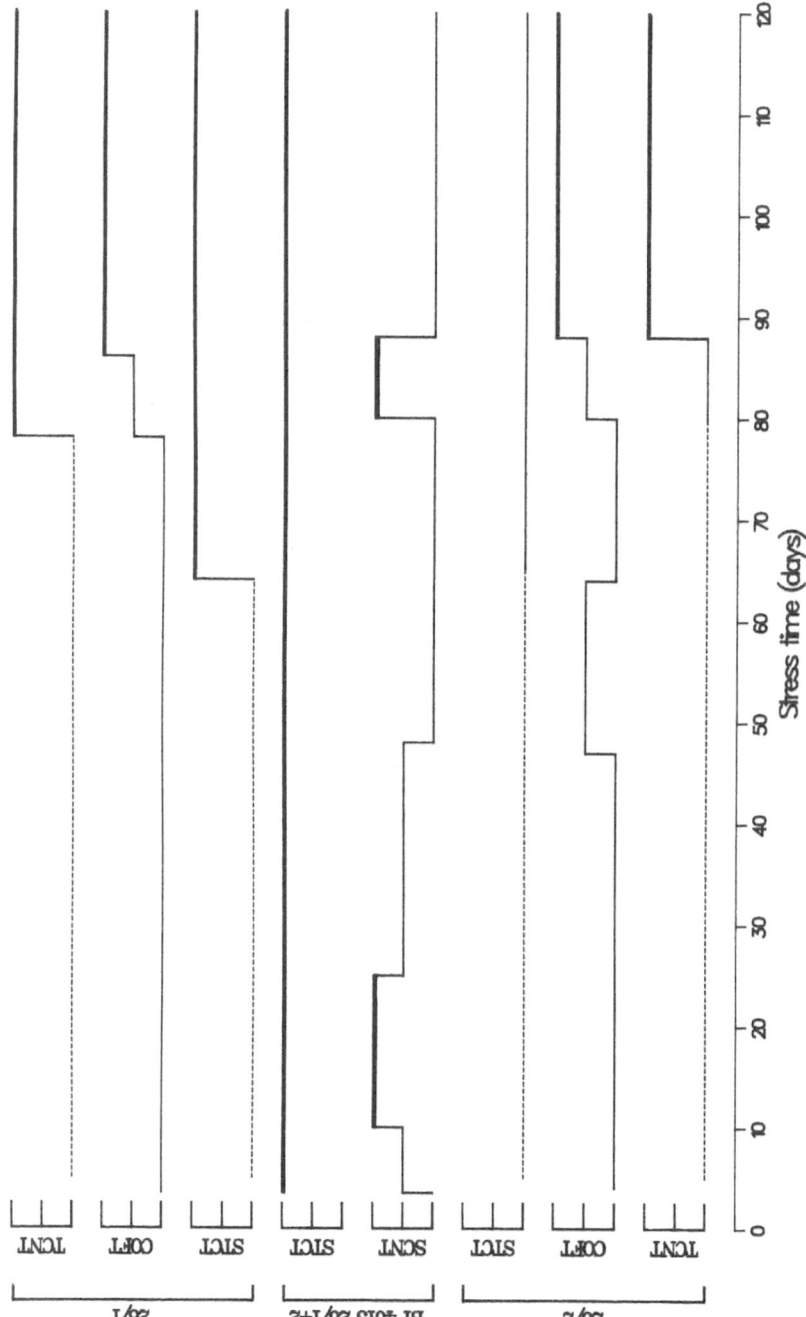

Fig. 4.37(a). Summary record of the detection of abnormality in devices of Batch 1 4013 26. The three levels from bottom to top represent normal, observal change and large or abnormal change in the test parameter. The central two tests are StCT and ScNT on the whole dual chip but with all inputs at one specific state. The other records are full tests on each dual circuit separately.

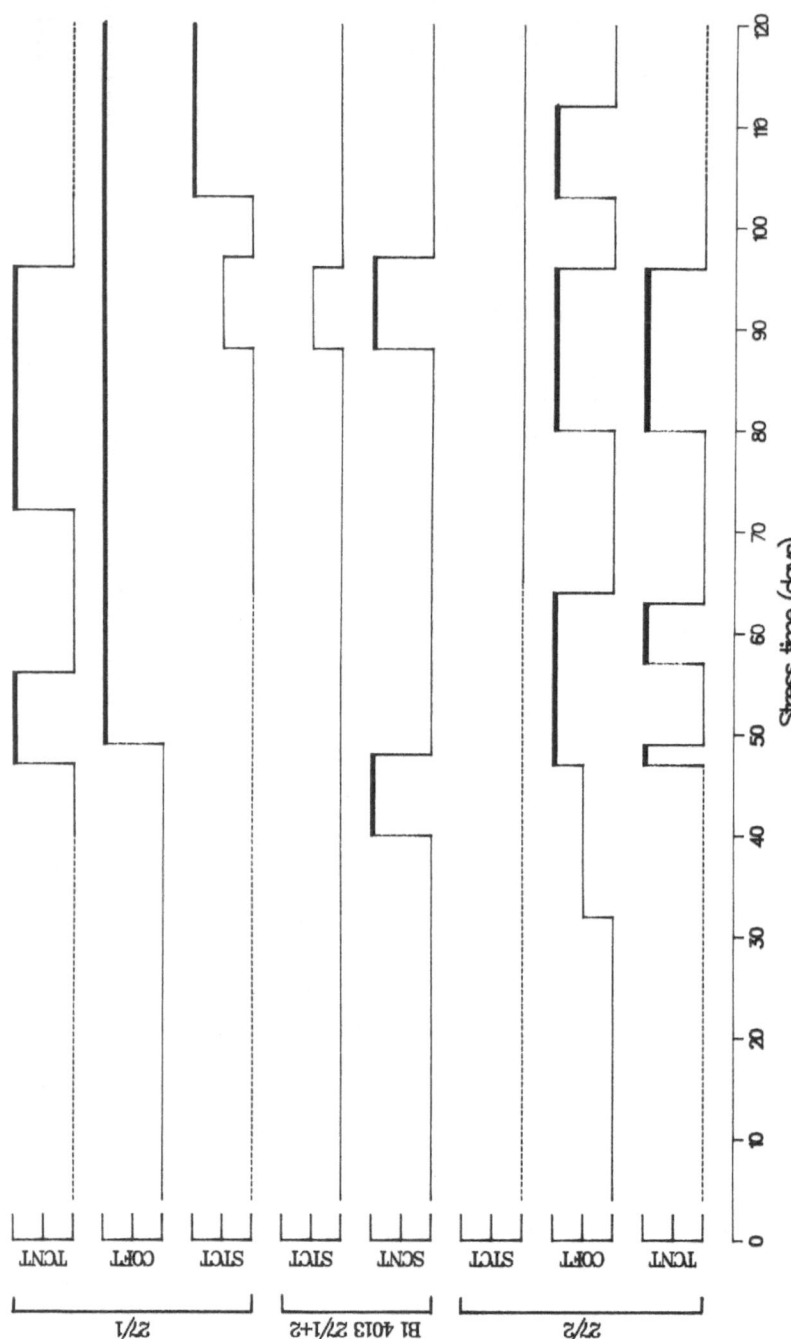

Fig. 4.37(b). Summary record of the detection of abnormality in devices of Batch 1 4013 27. The three levels from bottom to top represent normal, observal change and large or abnormal change in the test parameter. The central two tests are StCT and ScNT on the whole dual chip but with all inputs at one specific state. The other records are full tests on each dual circuit separately.

144

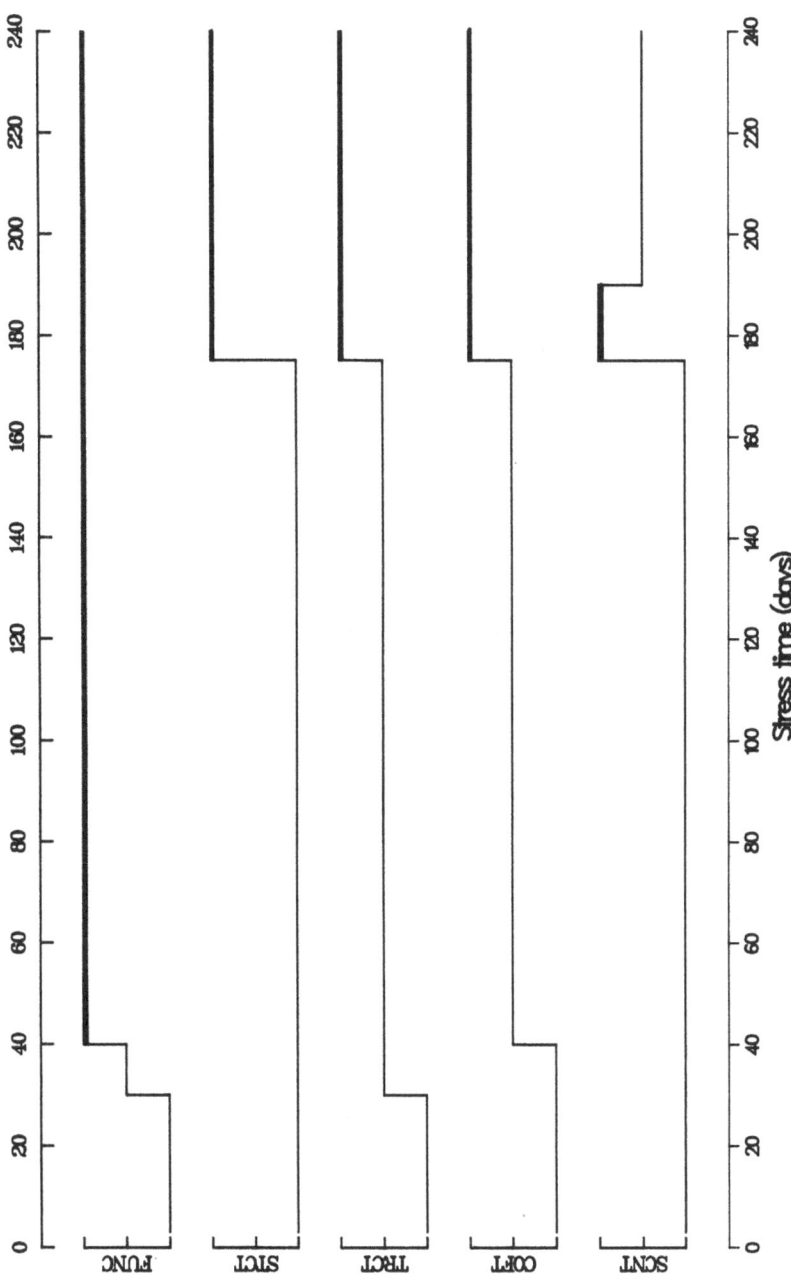

Fig. 4.37(c). Summary record of the detection of abnormality in devices of Batch 1 4013 28/1. The three levels from bottom to top represent normal, observal change and large or abnormal change in the test parameter.

145

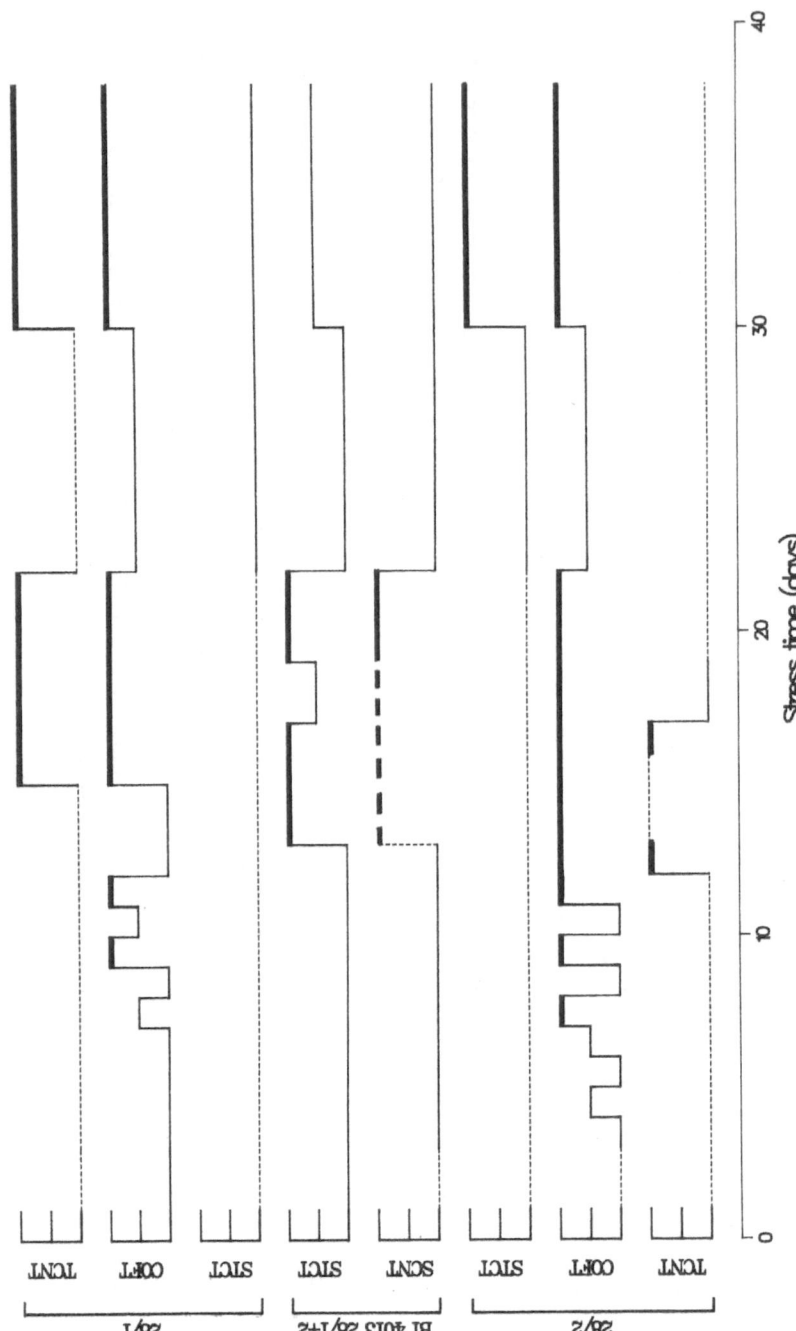

Fig. 4.37(d). Summary record of the detection of abnormality in devices of Batch 1 4013 29. The three levels from bottom to top represent normal, observal change and large or abnormal change in the test parameter. The central two tests are StCT an d ScNT on the whole dual chip but with all inputs at one specific state. The other records are full tests on each dual circuit separately.

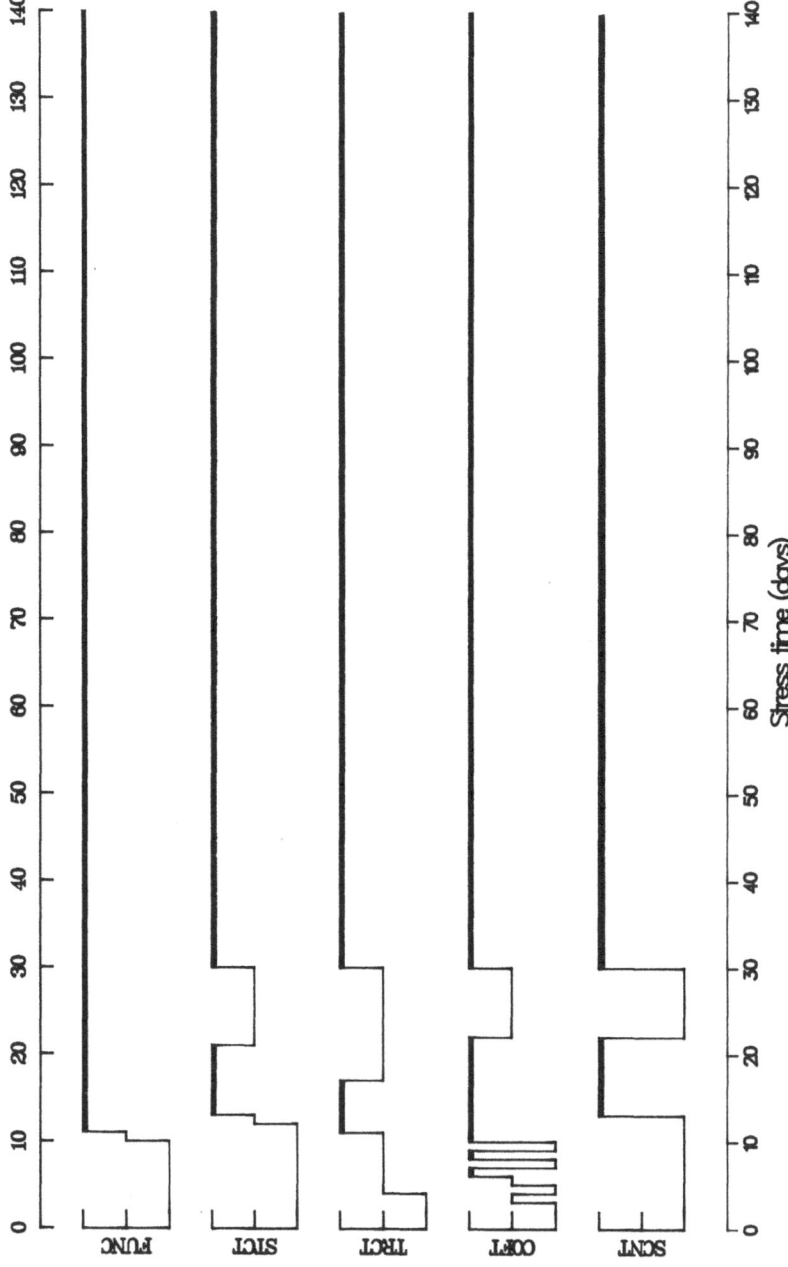

Fig. 4.37(e). Summary record of the detection of abnormality in devices of Batch 1 4013 29/2. The three levels from bottom to top represent normal, observal change and large or abnormal change in the test parameter.

147

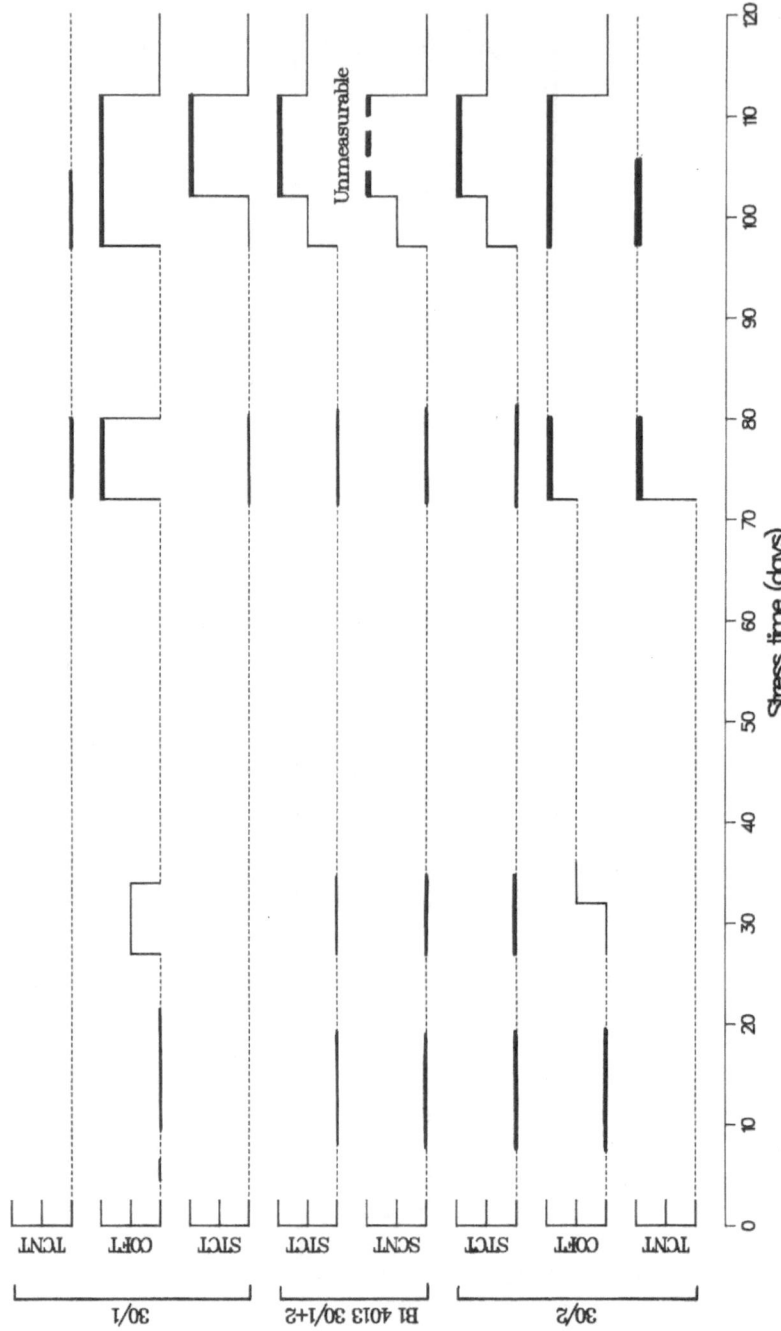

Fig. 4.37(f). Summary record of the detection of abnormality in devices of Batch 1 4013 30. The three levels from bottom to top represent normal, observal change and large or abnormal change in the test parameter. The central two tests are StCT and ScNT on the whole dual chip but with all inputs at one specific state. The other records are full tests on each dual circuit separately.

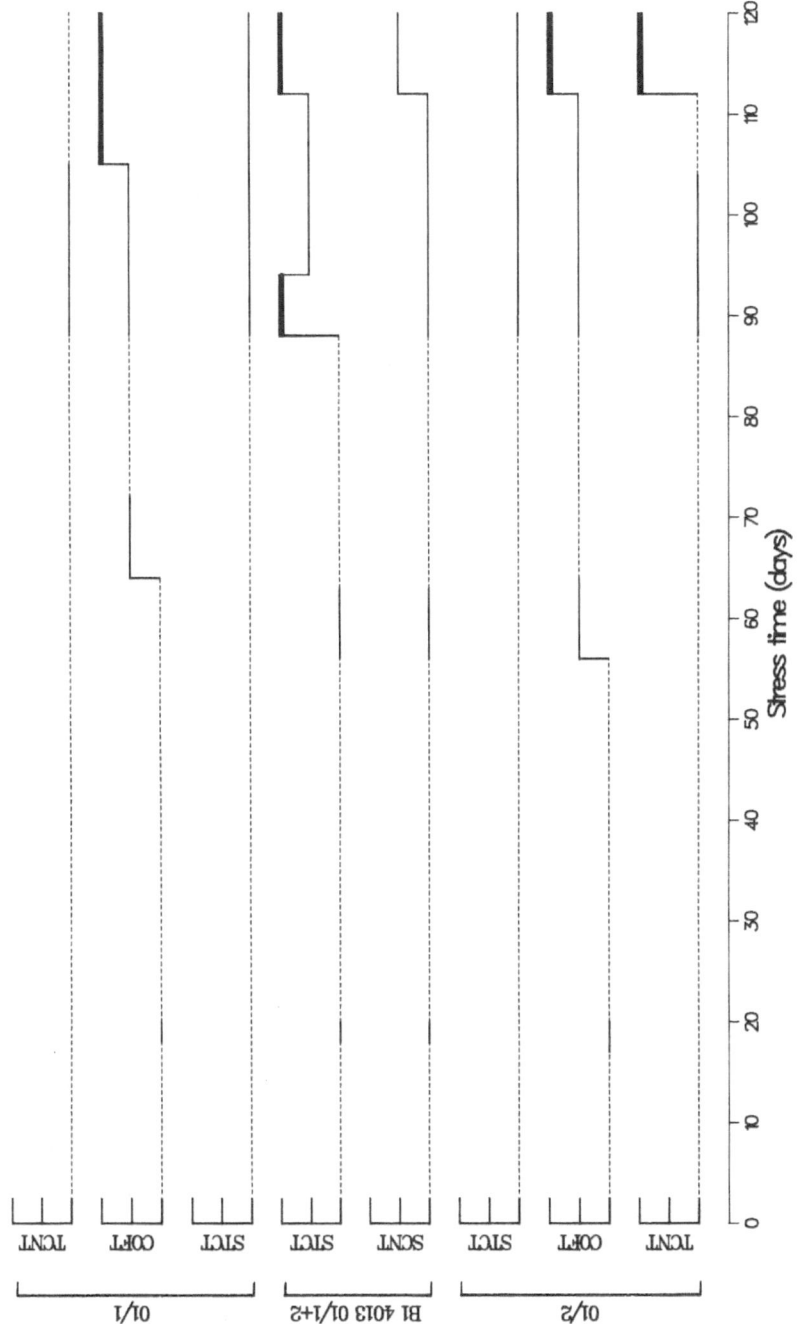

Fig. 4.37(g). Summmary record of the detection of abnormality in devices of Batch 1 4013 01. The three levels from bottom to top represent normal, observal change and large or abnormal change in the test parameter. The central two tests are StCT and ScNT on the whole dual chip but with all inputs at one specific state. The other records are full tests on each dual circuit separately.

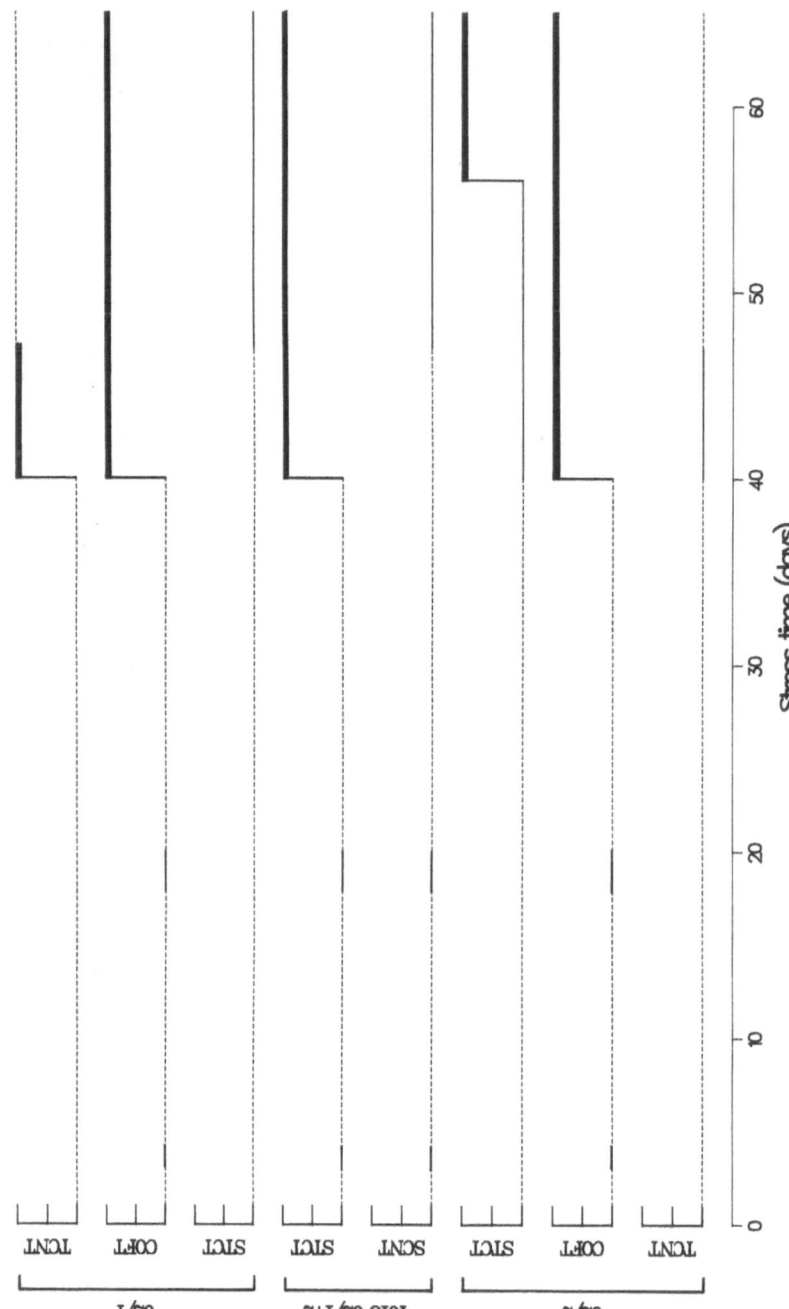

Fig. 4.37(h). Summary record of the detection of abnormality in devices of Batch 1 4013 02. The three levels from bottom to top represent normal, observal change and large or abnormal change in the test parameter. The central two tests are StCT and ScNT on the whole dual chip but with all inputs at one specific state. The other records are full tests on each dual circuit separately.

apparent fluctuation of the degree of abnormality is thus an extra indication of a defective device rather than a suggestion that the device is better than it had appeared.

Another feature is that both halves of the dual device eventually show abnormality, although one circuit usually indicates a problem in advance of the other. This is largely an artefact of this particular simple presentation. More detailed analysis of the tests show that usually the defect is in only one of the circuits. This will be discussed in a little more detail in Section 4.3.2 although a full discussion cannot be given here.

The cummulative results for Batch 1 over 240 days of stress are given in Figure 4.38. The results are given for the CoFT and StCT and also for the functional tests during the stress (FUNC). The devices are exercised during the stress to provide a realistic stress. The clocking frequency is about 1 kHz. A monitor circuit checks that the device still functions correctly. The FUNC plot shows those devices that cease to operate correctly at the elevated stress temperature, although nearly all the devices still function within specification at room temperature and all the other tests can be performed. It is seen that this 'test' is as good a predictor of degradation and potential failure as are any of the other tests. We therefore will include it in our analysis as another 'test'. It has great similarity to the 'validation temperature' which is sometimes used as a figure of merit. This is the maximum temperature at which a device will operate to specification. It thus has similarity to the CoFT since the temperature, supply voltage and frequency can each be varied to produce a malfunction condition.

4.3.2. Fault Location and Cause

The brief account given in Section 4.3.1 is greatly simplified and omits the considerable diagnostic detail contained in the full CoFT and StCT which investigate different parts of the circuit. A summary of the results obtained from these detailed tests concerning the location and type of fault is given here. The devices have not been subject to visual inspection but little is expected to be observed since the faults are probably purely electrical resulting from increased surface state density or oxide charge and are not the result of catastrophic failure which is readily observable by visual inspection. The devices still function. It may be that the electrical fault has been caused by a visible and obvious fabrication defect such as a dirt particle, but proof of the cause would be difficult to establish with so few samples.

The detailed analysis suggests that device B1 4013 26/2 [Figure 4.37(a)] has a defect in 26/1 which produced a breakdown of the p–n junction in one of the transistors connnected to Nodes D and F. The nodes are indicated in the diagram of Figure 3.17. There is evidence here and elsewhere that this defect caused the apparent degradation of 26/2.

Device B1 4013 27 [Figure 4.37(b)] developed two leakage faults in 27/1 at Nodes B and D and later also at Nodes C and D. These also account for the

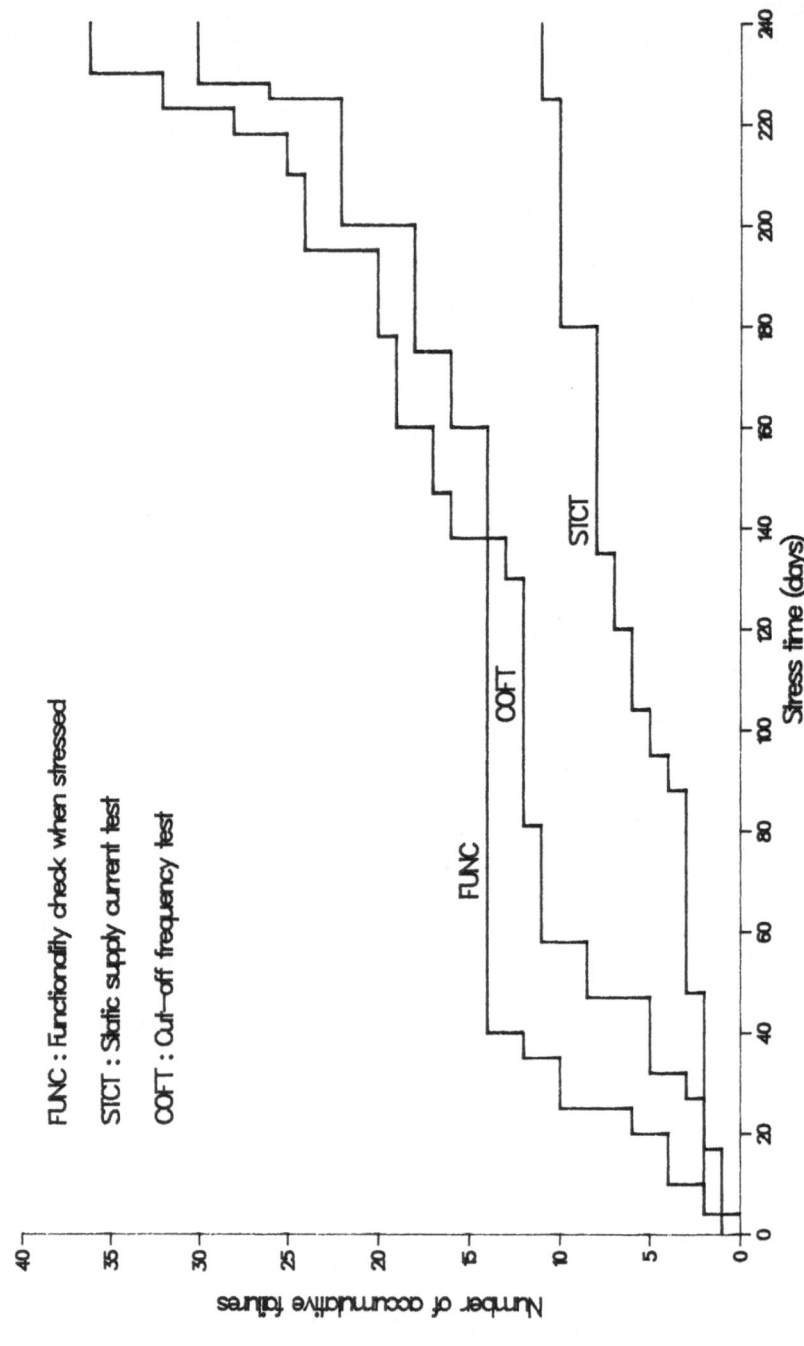

FUNC : Functionally check when stressed

STCT : Static supply current test

COFT : Cut-off frequency test

Fig. 4.38. Cumulative record of the detection of abnormality in devices of **Batch 1**. FUNC indicates a loss of functionality during the slow exercise during the temperature stress. The devices function at room temperature.

152

decrease in the cut-off frequency.

Device B1 4013 28/1 [Figure 4.37(c)] was not investigated for fault location.

Device B1 4013 29 [Figures 4.37(d) and (e)] showed several different faults which affected both 29/1 and 29/2.

Device B1 4013 30 [Figure 4.37(f)] developed a leakage fault in 30/1 located at Nodes B and D.

Device B1 4013 01 [Figure 4.37(g)] showed a very slow change in the cut-off frequency but the location of the defect has not been determined.

Device B1 4013 02 [Figure 4.37(h)] developed a leaky fault in 02/2 with a rather complex pattern which has not been analyzed.

4.3.3. Sensitivity of the Tests

The obvious question about these results is 'Which is the most sensitive or *best* test?' This is not an easy question to answer and it is not clear that there is a definitive answer.

The tests are each sensitive to changes in different device properties and will produce different electrical changes. Since the tests are required to detect any likely defect, several tests should be made together so that the detection of all failure modes can be detected. A discussion of the relative quality of the tests based on theory and practicability is given in the next section but here we will discuss the results of the experiments.

It should be remembered that these results are based on a single batch of a single device type under a single stress. Also the defects developed by this stress may not be typical of those existing immediately after fabrication due to imperfect processing.

The results of the outcome of the stress of Batch 1 have been presented, and other devices also studied but not discussed in detail show that all the tests presented in this report have value in the ability to predict a later degradation and probable failure. Other tests have been suggested, investigated and rejected. Variants of the tests used here also exist and have been mentioned. Examples are the validation temperature and the marginal voltage test which is the cut-off frequency test made by fixing the test frequency and reducing the supply voltage. These tests have their value but we considered them of less generality or more difficult to implement.

The tests with most discrimination and dynamic range are the StCT and CoFT which can investigate large or small subdivisions of the circuit and can also give a measure of the test over a wide range. They are also easier to implement than some other tests and this will be the main topic discussed in Chapter 5.

4.3.4. Comparison of the Tests

There are several aspects which will govern the choice of test: sensitivity, range of defect detected, speed, convenience, cost and ease of implementation. The detailed industrial implementation will be discussed in more detail in Chapter 5.

The ThVT is a good and simple test for assessing the quality of the device by an on-chip measurement of the threshold voltages and gain factors of all the input CMOS transistors. With several input terminals round the chip this test provides good statistical information about the quality and uniformity of the processing of the device. However, this test is likely to be of more use for process monitoring and quality control for yield purposes than for reliability analysis. This test is not able to detect any property in the chip beyond the input stage. We therefore do not consider it to be a general reliability test.

The two noise tests, StNT and TrNT, are early and sensitive indicators of possible defects. It is possible that they are able to detect defects that other tests cannot detect. However, we have no evidence of this since our range of stress and defects was probably restricted. It is likely that noise tests could detect electromigration failure or constrictions in the metallization before any of the other tests. The only other tests likely to detect these defects are the CoFT or delay time measurement of the TrCT, but the simulations indicated that the interconnect resistance would have to reach a very high value to reduce the current flow ability of the transistors. The sensitivity of the noise test could be very high if the StNT was performed for each input test vector and a full TrNT was performed on all the peaks of the current transient. In this way small subsections of the circuit can be investigated.

The drawbacks of the two noise tests are that they investigate a low frequency noise which is easily visible at frequencies of the order of Hertz so that a simple measurement takes a few seconds. This can be overcome as discussed in Chapter 5. Of more importance, the equipment and techniques are not simple or common so that industrial implementation may be difficult. The test has little significance in many applications.

The StCT has been seen to be extremely sensitive to the defects actually observed, it has a very large dynamic range and can give considerable diagnostic information. It is very versatile and, if the test vectors are chosen suitably, can either be a quick, insensitive test looking at large sections of the circuit or slow but detailed if many separate subsections are investigated individually. This appears to be an ideal test which can be implemented easily on modified industrial digital testers.

The CoFT and the TrCT used in the delay time mode give very similar information about the charging time of individual nodes. This has been found to be of great value in detecting defects. The former test has various variants depending on the supply voltage and temperature at which it is performed. With a low supply voltage it is more sensitive to threshold voltage changes, and at high temperatures it is more sensitive to leakage current changes.

The CoFT is easy to implement on modified digital testers and, by choice of the signal route, can give high discrimination between different subsections of a circuit. This test gives little diagnostic information.

The TrCT gives a very large amount of information. As a comparison with the CoFT we have concentrated on the pulse widths as a measure of the node capacitance charging rate. However, the detailed pulse shape and relative magnitudes and times of the subpulses within each pulse can provide a large amount of diagnostic information. We see this as a potentially powerful test, especially as the same measurements could be analyzed to give the TrNT data. However, analysis of the information obtained is not simple and the equipment needed is expensive and of high specification. High voltage resolution sampling is needed at a speed very much faster than the highest clock rate of the circuit being investigated.

The FUNC test is potentially very valuable . It is easy to implement and is very sensitive. As described so far it is of no value as a reliability test since the test is carried out during the destructive accelerated stress. However the basic principle can be applied with the suggestion that all the above tests be applied at an elevated temperature. Since burn-in is a valuable and established process, it is suggested that the test be carried out at elevated temperature during the burn-in period which is to some extent 'free-time'. The speed of the tests is very much shorter than the burn-in period so some of the test could be simplified to occupy more time. At burn-in temperature the device works less well, the effect of many defects becomes stronger and the test become easier to implement because the static current is larger and the delay times and charging rates are larger while the cut-off frequency is lower.

Experimentally our tests have detected mainly extra leakage currents. This may be a special feature of our methodology but many failure modes will produce extra currents. It is important that any set of tests chosen does not exclude detection of other failure modes such as open circuit due to electromigration.

It is perhaps also a symptom of our experiments that we see a very high correlation between the detection of faults by the StCT and the CoFT. Both tests can detect leakage currents and these may have been the dominant failure mode of our devices.

REFERENCES

Noise Test

Roder, V.H., 1979, *Frequenz*, **33**: 101-5.

Leakage Current Test

Dishman, J.M., Haszko, S.E., Marcus, R.B., Muraka, S.P., and Shang, T.T., 1979, *J. Appl. Phys.*, **50**: 2689-95.
Lycondes, N.E., and Childers, C.C., 1980, *IEEE Trans. Reliability*, **R-29**: 237-48.

Ravi, K.V., and Varker, C.J., 1974, *J. Appl. Phys.*, **45**: 263-71.

Varker, C.J., and Ravi, K.V., 1974, *J. Appl. Phys.*, **45**: 272-87.

Radiation Effects

Dressendorfer, P.V., Soden, J.M., Harrington, J.J., and Nordstrom, T.V., 1981, *IEEE Trans. Nucl. Sci.*, **NS-28**: 4281-7.

Chapter 5

IMPLEMENTATION OF THE TESTS
FOR INDUSTRIAL USE

5.1. INTRODUCTION

In the preceding chapters a series of analog tests have been described which can
be performed on digital ICs to assess their quality of manufacture and hence
their probable lifetime in use. The description has concentrated on the tests
themselves, the way in which they are performed, their sensitivity, their ability
to detect weak devices and a justification for supposing that the test results
actually produce a measure of the quality of the device. The development of the
tests, their assessment and their verification in sample batches of stressed devices
has been carried out under laboratory conditions such that the basis of the tests
is fully understood. The results suggested strongly that for the CMOS family
studied, the tests can be a valuable set of laboratory procedures.

In this chapter we discuss the value of the tests for industrial use and how the
results obtained here may be used to devise suitable test strategies for industrial
implementation. The needs of industry are very varied. At one extreme, for
fabrication yield assessment or for non-destructive failure location and failure
mode analysis the use of the tests needs to be flexible and detailed since a small
number of a wide variety of devices will be studied. A system similar to that
presented earlier would be suitable since the use is essentially in a laboratory
environment. At the other extreme, simplified tests may be suitable for routine
screening of all devices produced for military or high quality commercial category
devices so that a well founded reliability qualification may be given. This is
likely to be only justified on devices which are given a burn-in procedure. At an
intermediate level, a fairly full test program could be implemented for the small
number production of devices for extremely high reliability applications. Within
this range of requirements there will be compromises between the time taken
for the testing against the sensitivity of the detection level, and also between
the complexity, cost and versatility of the test equipment against the value of
detecting the weak devices. Ideally, fast tests on existing test equipment are
sought, although minor modifications may be acceptable.

In the following sections we discuss the best strategies for implementing
analog tests for specific purposes, the extension of the results to larger scale

ICs and the modifications needed for circuits involving both combinational and sequential circuits. A description is then given of the way some tests have been implemented on a specific digital circuit tester to suggest how other standard functional testers may be used. The final sections discuss how the tests could be implemented more quickly and more reliably using custom-made equipment rather than modified digital functional tests, and consider whether the tests are of value for other logic families and for analog integrated circuits.

5.2. TEST STRATEGIES

The basic minimum requirement for the quality of a digital IC is a full functional test. For the full set and sequence of input signal vectors this can take a long time even when carried out at the full speed of a circuit tester. Reduced test patterns which exercise each circuit node can be produced to speed up the process. Further reduced test patterns which do not carry out a complete test are also possible but with the natural loss of confidence.

Some of the analog tests described here need not take as much time as does the functional test so that for some purposes these could be considered as a preliminary screen for the full functional test.

With little modification the tests described can be used in the laboratory environment of a quality assessment or research activity. The detailed tests are non-destructive and can provide information on the precise location of a weak component and often some idea about the type of fault in the device. Armed with this information, further investigation by electron microscopy or other detailed methods can be concentrated on the weak component. Often an investigation of a circuit which has not failed catastrophically is difficult since there may be several blemishes of different types visible with most having no effect on the circuit. A non-destructive analysis before a destructive investigation involving de-encapsulation and selective etching can be particularly valuable. The tests CoFT, StCT, ScNT, TrCT and TrNT can all be used to locate the part of the circuit with the weak component. The noise tests, ScNT and TrNT, are slower and probably give no more information than their parent tests, the StCT and TrCT. As well as fault location, the extensions of the StCT with the full I~V curves and the full pattern of the TrCT can be used to diagnose the mode of the failure or weakness. Although it is less complete in its sampling over the whole circuit, the ThVT can be valuable in assessing the local threshold voltages at specific sites on the chip and the variation of these threshold voltages over the chip. Athough this has not been demonstrated specifically, this test could be extended to an investigation of the presence of mobile ions in the oxide using a bias-heat treatment method. The threshold voltages are measured after a series of moderate heat treatments at different selected bias levels so that the mobile ions are moved around the chip.

The use of the tests for fault location and diagnosis is valuable as an ajunct to the processing operation in order to determine the reason for low yield, large burn-in failure or short lifetime. Similarly, the outgoing and incoming quality assessment departments may need such techniques. The techniques are probably of greatest use in the failure analysis operation, although the work there is usually on totally failed devices.

The main use for the tests is as a predictor of quality and hence reliability. It has been explained earlier that the tests are not absolute tests and the values obtained as the result of the test depend on the process technology and detailed fabrication conditions. However, for a given design and process the tests can give a measure of quality between the members of a batch of devices which are nominally made in the same way. In general one can say that the highest quality devices will have lowest leakage current, and so on. The tests can therefore best be used as a method of sifting a large batch to select the best devices. Such tests could be used on a purely relative scale to separate devices to place into commercial, military or high reliability categories. Alternatively, or as an extension to the above, the selection may be made on an absolute scale. After many items of a specific circuit have been produced and tested, statistical information will become available about the spread of the test output values. If the statistical distribution of these values are established and confidence is obtained that the parameters are a measure of quality, then threshold criteria on one, or preferably several, tests can be produced to select good devices. This comparison of the result of tests on new devices with historical data can enable the processor to estimate whether the fabrication conditions are at the optimum values. Essentially the small fluctuations in each of the process operations has some effect on the test parameters and the device quality. These small variations may be used to suggest the direction of change needed in the process parameters to improve the device. The tests therefore have value as a production quality monitor to improve the yield.

For the selection of high reliability devices several tests should be made. We have seen that statistically all the tests selected show an ability to detect weakness and incipient failure. However the results produced are only on one variant of CMOS and two types of circuits so that general conclusions should be treated with some care. Of more importance, the detailed study of specific pathological devices shows that for some devices one test may be an early predictor while for other devices another test may prove better. This is to be expected since the failure modes of the devices are likely to be different. Thus to accommodate the wide variety of possible failure or weakness modes and mechanisms a wide selection of tests is necessary.

The basic compromises in the selection of the possible tests to use are between equipment cost, test time, test sensitivity and selectivity, ease of analysis, modification or extension of existing equipment and procedures. The optimum test strategy will be different for each industrial requirement and will depend on the existing operating system.

From the experience gained in the present set of comparison tests described earlier, several basic recommendations may be made. However these may not always be the best advice since individual circumstances can vary widely. The two excess noise tests are very valuable for detecting substandard devices but are rather specialist tests which require high quality, and not very common, equipment and techniques. An electrically quiet environment is also needed. The noise tests are also slow. The noise intensity varies as $1/f$ so that the lower frequencies give a larger signal. Typical frequencies at which the signal is above the measurement system noise is 1 Hz $\sim$ 10 Hz so that the test may take several seconds. Some signal averaging may be necessary to increase the signal-to-noise ratio and this will increase the time taken for the measurement. If this time was necessary for the measurement of the noise in each subsection of the circuit, defined perhaps by the input test vector as in the StCT, then the total measurement time would be extremely long. However, sampling and multiplexing techniques outlined later in Section 5.6.2 can prevent the total time increasing much beyond a few seconds, but with more circuit complexity. Since the noise results will also probably be of little use in diagnosing the failure mode, the two noise tests (ScNT and TrCT) are not recommended for general use.

The TrCT was found to be very informative, full of detail and rapidly gave similar information to that obtained in the CoFT, as well as additional information. The power of the technique cannot be underestimated in that the noise test, TrNT, can also be obtained with minor modifications and great detail is produced. It is, however, probably not suitable for routine, as opposed to laboratory, industrial use while other tests are available. In the method used in the present experiments the transient current through the circuit is recorded as different sections of the circuit are exercised. The current pulses are thus the result of the electrical properties of just these sections and hence give a large amount of precise information. In order to obtain the information, a measurement system bandwidth is needed which is greater than the highest frequency component within the circuit under test. For general applicability this bandwidth should be about ten times the expected frequency. In the measurement reported, the current flow signature was recorded using a sampling oscilloscope system. In this case the system bandwidth and the sample rate have to be above this bandwidth. In Section 5.6 simpler non-sampling techniques will be discussed. The speeds of digital ICs are progressively increasing so that any measurement system designed for one digital family will have but a short life before obsolescence. However, the problem is not insuperable since it is likely that sampling systems will always be faster than digital ICs. There will always be problems associated with the high frequencies, such as the need to match the impedances of the signal output connections to the recording system to prevent reflections. The high frequency transmission design within the IC is likely to be better than the output connections. Although powerful, this approach is also not recommended for general industrial use. In fact its power may not be an advantage since analysis of the information produced may not be easy. The easiest quantity to extract from the transient peaks is the peak width which gives information on the gate speed. The same, or closely related, information is available from the CoFT. The remaining tests are easy to implement and can give valuable information. The ThVT is simple and fast but gives only information sampled over the chip at the location of the input stages.

The CoFT can be performed on modified digital test machines, and can give some information about the location of the weak component but little diagnostic information. It has already been shown that this test essentially measures the propagation delay through each section of the circuit and thus gives very similar information to that supplied by the transient current pulse width in the TrCT. It is apparent, therefore, that this test could also be called a delay time test and in this form an implementation will be described later. A measurement is made of the propagation time delay through a particular path through the circuit defined by the appropriate set of test vector sequences.

The StCT is most valuable since it is easy to implement on modified digital test equipment and can give weakness location information and, from the full I~V curve, considerable information about the failure mode and mechanism. This test is very flexible and allows freedom for the operator to carry out a fast, coarse test or a slow, sensitive test. If we assume that we have sufficient analytical power so that we can devise input test vectors and test vector sequences so that we can turn internal transistors and gates ON or OFF at will then we can devise different test strategies. At one extreme we can apply two tests, one with half the transistors ON and the others OFF and the other the complementary state. At the other extreme we can apply a number of input test vectors corresponding to the number of transistors so that in each test one transistor is ON and all the others are OFF. The first test procedure is fast but will not be sensitive since the actual current measurement sensitivity will be finite so that only one extremely leaky transistor can be detected among the normal leakage current sum of all the transistors. This coarse test will only be able to discriminate very bad devices. At the other extreme the system will be able to detect very small deviations from the ideal for each transistor when it is tested. Information theory suggests that suitable optimum test strategies can be devised so that at any time a discrimination is made between equal numbers of ON devices. In practice, the difficulty of devising an optimum set of test vectors is great until automatic analysis techniques are devised to determine the test vector for the desired internal node logic state. One method of reducing the difficulty of the analysis of the internal logic state is to perform a differential measurement whereby each device is measured and the results compared to a similar reference device. Any large, non- systematic departure from equality would indicate a fault. The comparison could be made in analog form in real time or by comparison of one set of measurements with the record of the measurements on the comparison device.

The experimental test which was seen in Chapter 4 to provide the earliest prediction of failure was the functional test performed by the exercise curcuit during the thermal stress. This was performed at a low frequency, 1 kHz. This test is, in effect, similar to the CoFT but performed at a high temperature. It is worth comparing the various tests of this type. They all detect some criterion for malfunction by altering the power supply voltage, operation frequency or operating temperature. An ideal experiment would construct the three-dimensional $(V_{dd}, f, T,)$ surface of failure.

The CoFT described in this book measures the frequency to malfunction at a standard temperature near room temperature. The variation of this frequency with V_{dd} shown in Figure 3.20 indicates a smooth variation in this maximum frequency down to a minimum V_{dd} approximately equal to the sum of the threshold voltages of the n- and p-channel transistors. To obtain a good sensitivity for changes in the threshold voltages and gain factors, V_{dd} and K, the measurements were made at a low voltage but far enough above the minimum voltage that all devices would always operate until very near total failure. The 'marginal voltage test' again operates at a standard temperature near room temperature but measures the value of V_{dd} for malfunction at a fixed, high frequency near the specification maximum frequency. This test is sensitive mainly to the threshold voltage value rather than the gain factor.

Another related test is the measurement of the 'validation temperature'. The device temperature is raised to the temperature of malfunction at the specification V_{dd} and frequency. As the temperature is raised, the leakage currents increase and the gain factor decreases as the carrier mobility decreases until the actual maximum operating frequency is reduced to the specification limit frequency. This test is thus sensitive mainly to the gain factor and leakage change. This test is very inconvenient since the device temperature has to be altered in a controlled and measurable way. It is inherently slow because of the long thermal time constant of the device.

The measurement that we performed by observing the malfunction at a fixed high temperature and fixed low frequency is thus another similar measurement. It is likely to be sensitive mainly to leakage currents, although changes in all the parameters would be detected. This test is not recommended since it is carried out at a temperature sufficient for thermal stress and is hence not non-destructive.

We recommend that the tests that we have observed to be of value and are easy to implement are carried out at an elevated but not very high temperature in order to accentuate all the deleterious effects in the device. Rather than introduce a new process, it is suggested that the tests are performed during the burn-in period which at present is only occupied by an exercising process. No extra time is then needed for the tests and the only extra cost is the test equipment and the extra complexity of the burn-in chambers.

During the burn-in process the tests suggested are: the cut-off frequency test, CoFT; the threshold voltage test, ThVT; and the static current leakage test, StCT. There are added advantages in performing these tests at the burn-in temperature. The leakage currents are much larger (typically in the simulation they are 48 μA at 125^0C for the 4013) and increase by about a factor of two for each 10^0C rise in temperature. The peak width of the TrCP increases (typically showing a 45% increase at 125°C) so that the tests are easier to perform. There seems no obvious reason why standard digital test equipment, modified as described later in Section 5.4 or 5.5, could not be used.

5.3. EXTENSION OF TESTS TO VLSI

The method and results of the tests given in Chapters 3 and 4 show that the tests are valuable for reliability analysis and that much is known about the information is available from the tests. The extension of the fundamental result to industrial implementation requires the extension to larger scale ICs with more transistors and the design of faster, simpler and more cost-effective equipment and procedures so that a large number of devices may be tested rapidly.

The two main electrical tests considered for routine use are the cut-off frequency test, CoFT, and the static current test, StCT. As the scale of the circuit increases, the CoFT becomes a much longer test since the number of current paths through the circuit, from input to output, which need to be tested increases significantly. Depending on the structure and complexity of the circuit, some advantage may be obtained by constructing a few long paths through many gates rather than many short paths which each test a few gates. However, this reduction in the time taken is at the expense of discrimination or sensitivity of the test. However, the time taken for the StCT does not increase so rapidly with the size of the circuit since the test strategies may be adapted for rapid search. These were outlined in Section 5.2. Once a level of sensitivity has been decided, then subsections of the circuit are selected in turn which contain the maximum number of transistors within which one defective device can be detected at the required selectivity limit.

The total time taken for a test depends on the number of measurements made during the test and the time taken for each measurement. The aim of practical test equipment would be to have a test speed or clock rate only a little slower than a conventional functional tester. This aim may not be necessary in practice since only small numbers of devices are likely to be tested for high reliability, but many different device types could be expected so that the setting-up time for a new device may dominate the total test time.

For the CoFT each measurement will need to take a time for each signal path of several times the total transit time of the signal through the path. For the StCT the main time taken will be in the analog current measurement at each test vector.

The strategy for the selection of an optimum set of test vectors for each of these tests is very important. This optimum strategy has been widely studied for the functional test of digital circuits and has been discussed in Section 3.3 and is reviewed here.

Digital circuits are generally split into two types when considering testability: combinational and sequential. The development of a standard functional test set for a digital circuit is based on the assumptions that a defect will cause a detectable logical fault on one of the primary outputs and that every possible fault can be detected by developing a set of test vectors so that every node in the circuit is switched between both LOW to HIGH and HIGH to LOW. The test set is then normally subjected to techniques such as fault collapsing and cyclic redundancy checking for analyzing the output vectors.

In the context of reliability testing, the above philosophy causes some problems:

1. Fault blocking due to reconvergence, or other pathways blocking the propagation of the fault to a primary output, can prevent some gates being tested.
2. Redundancy is accepted as a by-product of using modular techniques in combinational logic design and the use of standard cells and macros in semi-custom ICs.
3. The functional testing approach by definition assumes a defect in the circuit will cause a stuck-at-1 stuck-at-0 fault at one or more of the internal nodes of the circuit. This is not necessarily the case.

Fault blocking is a major problem in testing VLSICs and is normally caused by redundancy. One solution to this problem is the incorporation of 'design for testability' techniques into the circuit design; however, in some cases this is impractical and a need to monitor the performance of these elements using a different technique may well be necessary. Gate array and cell-based designs will almost certainly contain a high degreee of redundancy and for these low cost semi-custom devices design for testability considerations may be impractical. Detection of degradation mechanisms which do not initially cause complete functional failure on one or more nodes cannot be achieved using a functional testing technique. New constraints on the test sets are therefore required to be compatible with these new techniques. The techniques suggested for use in LSI and VLSI devices are the static leakage current test and the cut-off frequency test. We shall consider the test set for both combinational and sequential circuits. The requirement of the test set for StCT is to switch every CMOS circuit node through both its logical states. This will be achieved by ensuring that each logical element is exercised through its full truth table. As a substantial stabilization period is required before the leakage current is measured, it is clear that the test set must contain the minimum number of test vectors. This can be achieved by fault collapsing techniques and closer analysis of the transistor level design of the circuit elements as in Section 3.3.1.2 shows that there are redundant conditions in the truth tables. We can therefore conclude that a *reduced* minimum functional test set is required for these tests.

The CoFT relies upon input vectors exercising logical pathways through the circuit to analyze the rise and fall times of the elements making up that path. If we consider that a deviant device may show a rise time of say 25% greater than its nominal value, then the total delay increase through 100 similar gates will be only 0.25%. Two important limitations to the test set are that each transistor in the circuit must be part of at least one path and that the chains must be kept as short as necessary for the required resolution and sensitivity.

The test set required for this test is therefore different from the functional test set discussed earlier and consequently different algorithms and analysis techniques for definition and measurement will be required.

To perform the delay test version of the CoFT on a combinational circuit we need to generate two reference clocks, one to sensitize the primary input with the test pattern, and one to sample the resulting pattern from the primary outputs. There have been several papers published on delay testing. Most have only been concerned with functional testing and have hence set the clock delays to values appropriate to each path under test, whereas others have been solely concerned with ensuring that all primary outputs are true during a specified clock delay. The procedure used for our reliability test will be to decrement the relative clock delays until we get a functional failure on one of the primary outputs. The test vector causing the failure is detected, the relative clock delay logged and the process repeated until all the test vectors have had their delays to failure logged. By deriving a suitable test set, each gate in the circuit can be monitored by analyzing the delay signatures.

This technique has the advantage that the test will only require a data generator capable of generating the full input test set and clocks, a data analyzer capable of resolving down to a delay equal to the smallest single element, and a central processor, all of which are used as standard functional test equipment. The time to conduct this delay test will be equal to

$$P + V \sum_{n=S}^{N} T_n,$$

where T_n = delay between reference clocks with values between
$\quad S$ = smallest delay and
$\quad N$ = largest delay,
$\quad P$ = processing time and
$\quad V$ = number of test vectors.

For the StCT let us first consider a standard CMOS element, a flip-flop as shown in Figure 2.2(a).

It is clear that we cannot test the transmission gate as it is not connected between the supply rails. We can, however, test the inverters which, as for the combinational case, need to be switched through both logic levels. For this simple structure our test set will consist of a series of 0s and 1s clocked through the circuit. However, for circuits with asynchronous elements and different clocking strategies the test set may become more complex.

We now consider the CoFT for sequential combinational logic. Sequential circuits are defined by the presence of some form of memory or clocked element within the circuit. Because the delay test relies on signal paths between the primary inputs and outputs independent of any external signals such as enable clock signals, it is clear that a different test must be employed.

The upper cut-off frequency test implemented as a delay-time test will in effect measure the delay of the slowest sequential element in a path which consists of elements of a shift-register, counter, and so on. To ensure that the cut-off frequency measured is in fact due to the circuit and not due to the measurement equipment the set-up time is first measured, ie, clock delay after data has become valid. For the test on the 4014 this delay was increased by 20 nanoseconds and used as the set clock delay for the cut-off frequency measurement.

Since a synchronous sequential circuit relies upon clock pulses to propagate a signal through the circuit, we cannot perform a delay test unless an output from each stage of the synchronous circuit is available. The cut-off frequency test will in effect measure the delay of the stage with the longest delay and, as this delay will depend upon whether the transition is low-high or high to low, the test set must exercise the circuit through all transitions. A test set of 110011001100 has been used on synchronous elements as this provides transitions between all possible states.

5.4. IMPLEMENTATION OF THE TEST ON A DIGITAL TESTER

5.4.1. The Semi-Custom Multiplier Chip

The next increase in scale of circuit was investigated using a semi-custom test chip consisting of a 4 × 4-bit multiplier followed by an 8-bit shift-register. This has been described briefly in Section 2.2 and sections are shown in Figure 5.1.

The choice of a test vehicle for furthering our reliability studies on larger scale devices has been complicated by a number of factors such as limitation due to equipment, reluctance of manufacturers to supply information on devices and difficulty in finding designs which would be easy to understand with our tests. While the tests are expected to be useful generally, at this stage in the investigation a simple test vehicle was needed. Because of the latter two difficulties, commercial devices were eliminated as possible test vehicles.

The decision to use the gate array fabricated as a 4 × 4 multiplier was made because: the devices are of comparatively recent technology; the devices are testable using equipment available for the project; the design is combinational and incorporate genuine test problems such as redundancy, delay path variations and multiple outputs; and the SPICE parameters and layout information are available so as to allow simulation work to be carried out without the need to reverse engineer the devices.

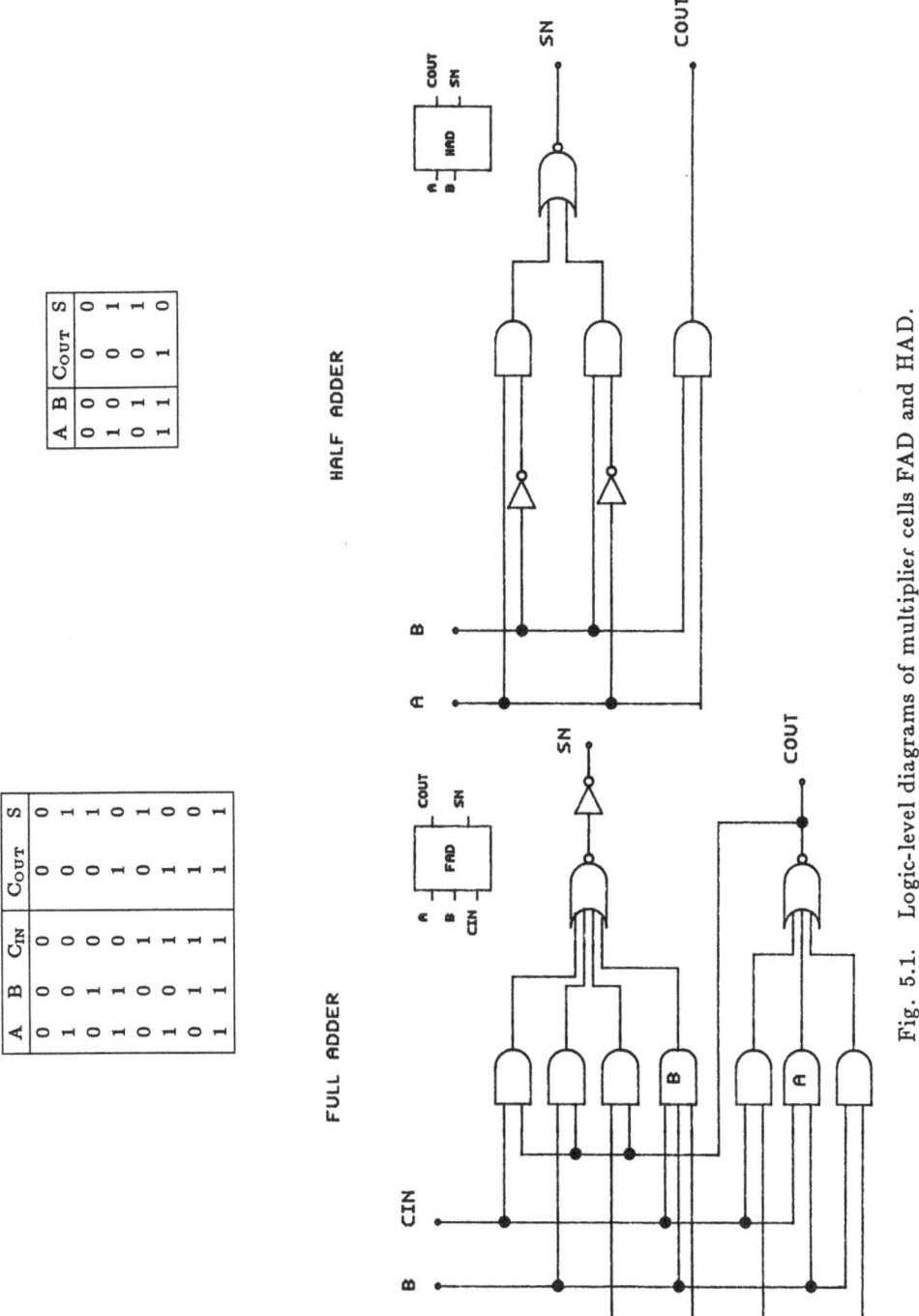

A	B	C_OUT	S
0	0	0	0
1	0	0	1
0	1	0	1
1	1	1	0

A	B	C_IN	C_OUT	S
0	0	0	0	0
1	0	0	0	1
0	1	0	0	1
1	1	0	1	0
0	0	1	0	1
1	0	1	1	0
0	1	1	1	0
1	1	1	1	1

HALF ADDER

FULL ADDER

Fig. 5.1. Logic-level diagrams of multiplier cells FAD and HAD.

167

Table 5.1. Test Set for the Multiplier Showing Both Input and Output Vectors. The Check Bit is the Output Which Must be Monitored for the Particular Input Vector.

Initial verification of zero condition
Inputs   00 10 20 40 80 01 02 04 08
Outputs 00 00 00 00 00 00 00 00 00
Functional Test

Vector no	Input vector	Output vector	Check bit	Vector no	Input vector	Output vector	Check bit
1	11	01	0	26	C6	48	6
2	11+c	02	1	27	84	20	5
3	00+c	01	0	28	CC	90	7
4	12	02	1	29	88	40	6
5	13*c	04	2	30	EC	A8	7
6	14	04	2	31	42	08	3
7	17	07	3	32	33	09	3
8	18	08	3	33	35	OF	4
9	1F+c	10	4	34	37	15	2
10	21	02	1	35	3B	21	5
11	31+c	04	1	36	3E	2A	3
12	22	04	2	37	75	23	5
13	36	12	4	38	67	2A	3
14	24	08	3	39	7A	46	6
15	3C	24	5	40	6E	42	4
16	28	10	4	41	7F+c	6A	5
17	3F+c	2E	5	–	–	–	–
18	41	04	2	43	E5	46	6
19	63	12	4	44	C7	54	4
20	66	24	5	45	EA	8C	7
21	44	10	4	46	CE	A8	5
22	6C	48	6	49	FE	D2	6
23	48	20	5	48	A5	32	4
23A	7C	54	6	49	55	19	3
24	81	08	3				
25	C3	24	4				
25A	82	10	4				

NB. Input vectors are in Hexadecimal.
    XY = IN8 IN7 IN6 IN5 IN4 IN3 IN2 IN1.
    Output vectors are in Hexadecimal.
    ZZ = OUT8 OUT7 OUT6 OUT5 OUT4 OUT3 OUT2 OUT1.
    OUTPUT = X*Y;   C = Carry high.

5.4.2. The Multiplier Test Set

As we need to perform both delay and quasi-static leakage current tests on the multiplier, we need to develop a test set suitable for both tests. The requirement of this test set is:

1. For the leakage current: that all p–n junctions are stressed and that the test set should be a minimum length.
2. For the delay test: that the test exercises every node in the circuit, that each node is part of at least one circuit path and that the paths are of minimum length.

As can be seen from Tables 5.1 and 5.2, test vectors have been devised to exercise each adder and gate through each state of its operation. The test routine is such that each vector is applied and then removed from the circuit by a reset condition of all zeros. This ensures that each node is driven HIGH to LOW as well as LOW to HIGH on each test.

With reference to Table 5.2, the primary test vectors are the vectors which will create the required condition whilst exercising the mimimum number of nodes in the rest of the circuit, whereas the secondary test vectors will set the required condition but exercise more nodes than the primary vectors. It can be seen that by deleting any primary test vector which can be replaced by one of its secondaries, the test set length can be reduced.

There are four redundant conditions within the multiplier structure in Figure 5.2. From Table 5.2 it can be seen that the Full Adders 5, 9 and 13 will never have inputs of A=0, B=1, C=1 and that Full Adder 5 also never has inputs of A=0, B=1, C=1. On closer analysis it can be seen that the output of gate A (seen on Figure 5.1) on adders 5, 9 and 13 and the output gate on adder 5 will never go high. This redundancy is to be expected due to the use of standard cells.

5.5.   RELIABILITY TEST EQUIPMENT AND TECHNIQUES

The techniques described below have been developed for combinational circuits with the multiplier structure having been used as an example. An attempt has been made to develop an automated test system for rapid reliability screeening of VLSI devices with equipment which is close to that used in industry.

The system is shown in Figure 5.3 and consists of an HP220 central computer as the main controller, an HP8080 data generator supplying the device with the input list patterns and the reference clocks, and an HP8182 data analyzer, both of which are programmable over the IEEE-488 bus. The supply for the devices is from the programmable power supply, the leakage current being monitored by the DVM via the current–voltage converter board. As leakage current is strongly

Table 5.2. Multiple Test Vectors

Element	A·B·C̄in	Ā·B·C̄in	A·B̄·C̄in	A·B̄·Cin	Ā·B·Cin	A·B·Cin	Ā·B̄·Cin
D 1	8 23A 47 / 35 15 / 39 36	7		9 12 / 4 / 41			
D 2	6 13 37 / 15 49 / 33 36	5	11 / 33	7 / 17 / 9 41			
D 3	4 35 39 / 36 13 / 47 32	2	11 / 33	5 / 17 / 7 41			
D 4	1 37 34 / 35 43 / 49 32	3		2 / 17 / 5	33 41 / 17 5 / 11 9		
D 5	16 39 22 / 40 30 / 45	9		17	15 35 36 / 47 / 23A	*	13 / 32 / 34
D 6	14 40 43 / 20 22 / 41 48	7	39	15 23A	13 33 / 34	36 47	32
D 7	12 20 38 / 39 17 / 45 19	5 / 15	37 6 / 49 / 15	13 47 / 36	32 35	34	11
D 8	10 38 43 / 48 19 / 17 41	4	39 13 / 36 / 47	11 32 33 / 35 / 34			
D 9	23 45 / 46 / 28	15	36 / 17 / 35	23A	22 40 / 30 / 39	41 47	20 / 37 / 38
D 10	21 49 28 / 43 26 / 44 46	9	45 35 / 33 / 16	22 30	20 / 37 47 / 38 / 41	40 39	19 37 / 20 / 38

State

| | | | | | State | | | |
Element	$A \cdot B \cdot \bar{C}_{in}$	$\bar{A} \cdot B \cdot \bar{C}_{in}$	$A \cdot \bar{B} \cdot \bar{C}_{in}$	$A \cdot B \cdot C_{in}$	$\bar{A} \cdot B \cdot C_{in}$	$A \cdot \bar{B} \cdot C_{in}$	$A \cdot B \cdot C_{in}$	$\bar{A} \cdot \bar{B} \cdot C_{in}$
D 11	42 46 18A 44 25 45 26	36 30 7 32 48 43 14	40 20 47 39	19	37	38	41	49
D 12	42 25 18 4 44 43	40 20 5 39 15 45 12	37 49 19 38 41					
D 13	29	39 23A 22 40 41	30	44 28 45 46	*		47	43 26 44
D 14	48 27 30	36 23 15 37 35 38 21	47 28	43 26 44	45	46		42
D 15	47 25A 25	33 19 9 28 16 40 49	46 26 45	42	43	44		48
D 16	24	26 14 7 22 17 36 32	42 48 25 43 44					

NB. Not all secondary vectors are included.
 * No test vector exists.

Format of Table: x y_1 / y_2 / y_3 x = primary vector; y = secondary vector.

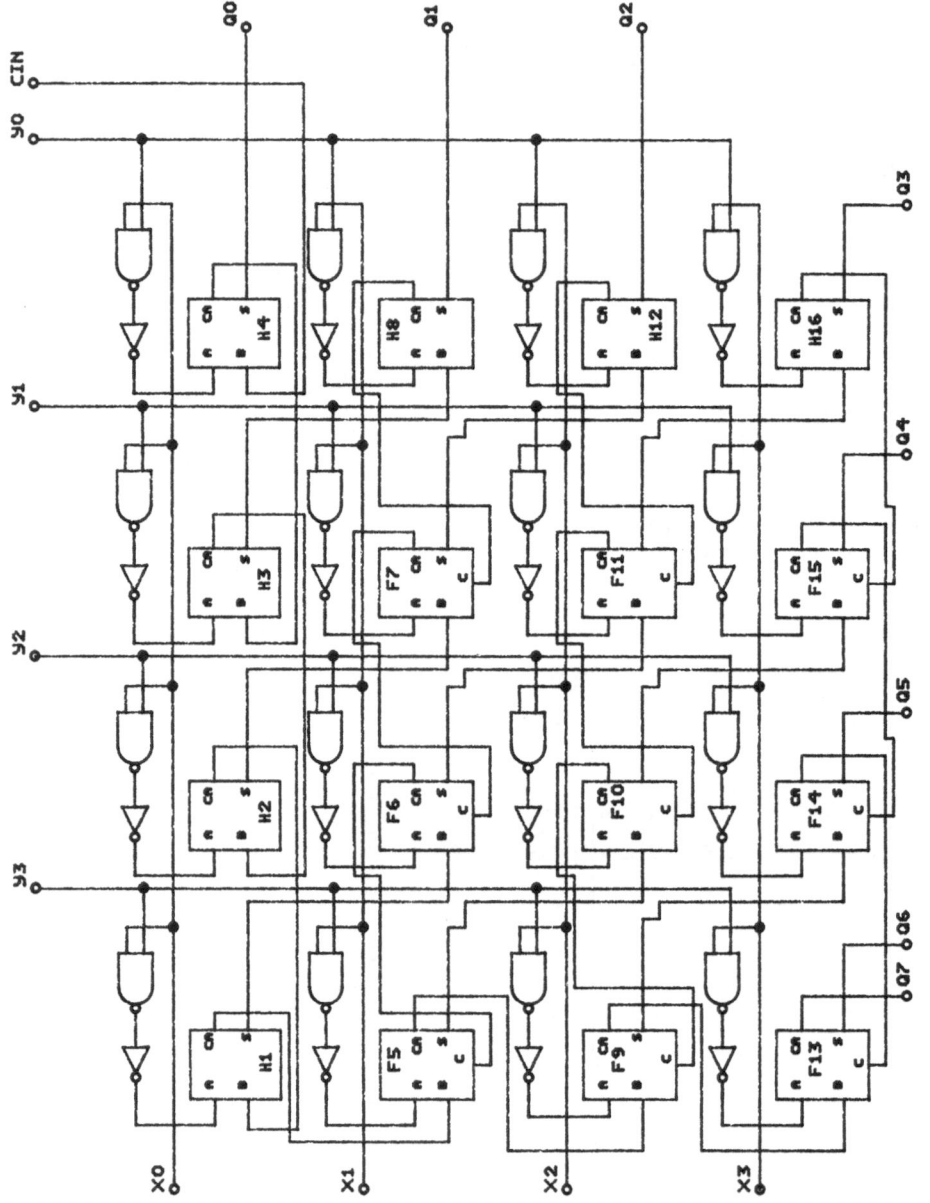

Fig. 5.2.   Schematic circuit of multiplier chip showing a typical propagation path through the network.

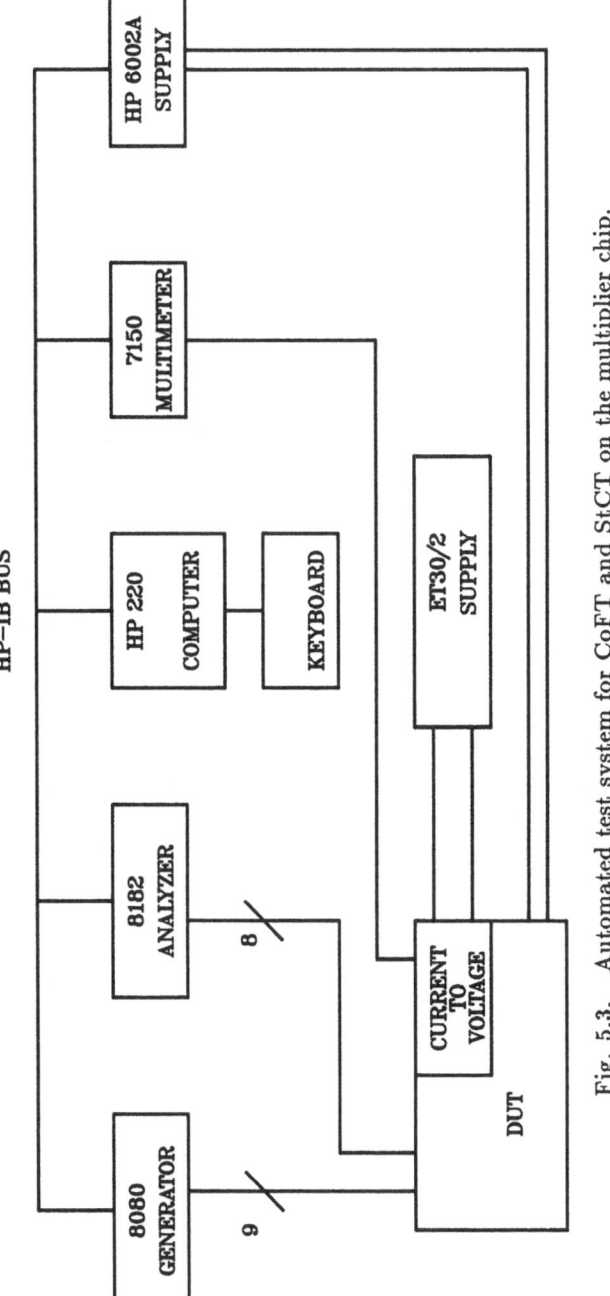

Fig. 5.3. Automated test system for CoFT and StCT on the multiplier chip.

dependent upon temperature, a control box was used to keep the temperature of the devices constant at 25 ± 1^0C.

The system, when initiated, is run under program control, the delay test being performed first. As no procedure for autogenerating the test sets required exists, they must be manually derived and keyed into the system interactively together with set up parameters such as supply voltage, initial delay times, delay documents and the test frequency.

Once the system has loaded the test set for the device under test, the delay test is initiated. A flow diagram for this test is shown in Figure 5.4. Once the supply has been set, test vectors are applied to the circuit at a frequency of 1 MHz on each positive edge of reference clock 1. After each test vector has been applied, the active output is sampled at a time interval after reference clock 1 and compared to the expected response held in memory. After the full test set has been applied, the delay is decremented and the test repeated.

Should an error be detected, the vector causing the error is deleted and the delay logged. The test is run until all test vectors have been deleted and logged.

On exit from the delay test, the system reloads the test set and initiates the leakage current test. The flow diagram for this test is shown in Figure 5.5. The supply and logic voltage levels are set at the maximum value specified by the data sheets.

The current-to-voltage converter is integrated into the system as follows. The V_{ss} line from the device is connected to ground via the input of the GE/RCA 8084 logarithmic current–voltage converter. Test vectors are applied to the circuit at a frequency of about 1 kHz so as to allow the output of the current–voltage converter to settle before it is sampled by the DVM. The control routine scales the output of the DVM to give a true value of current for storage. The logarithmic current–voltage converter is slow but can accomodate a very large dynamic range. For fast response over a smaller range (for example, for a pass/fail test on an established device) than a linear feedback current–voltage converter may be used.

Before the test system is initialized the current–voltage converter must be adjusted for offset and gain. This is achieved by temporarily connecting an external reference current to the input of the 8048 and adjusting the gain so as to obtain an output of 1 V per decade change in current. The offset is adjusted in the usual way by grounding the 8048 input.

Results are stored and analyzed as deviations from a standard reference derived from a known good device. Output plots can therefore be obtained of test vector–delay and test–vector leakage. This technique allows rapid analysis, a threshold of deviation is all that needs to be defined when the purpose of the test is to reject deviant devices. Should a closer analysis of the results be necessary for such exercises as failure analysis, these plots may be processed in more detail to pinpoint degraded areas and predict degradation mechanisms.

Figures 5.6 and 5.7 show typical results of the two tests. The tests produce a characteristic signature which for the delay time varies only slightly from a mean value, whereas the delay time-test vector signature has large variations between

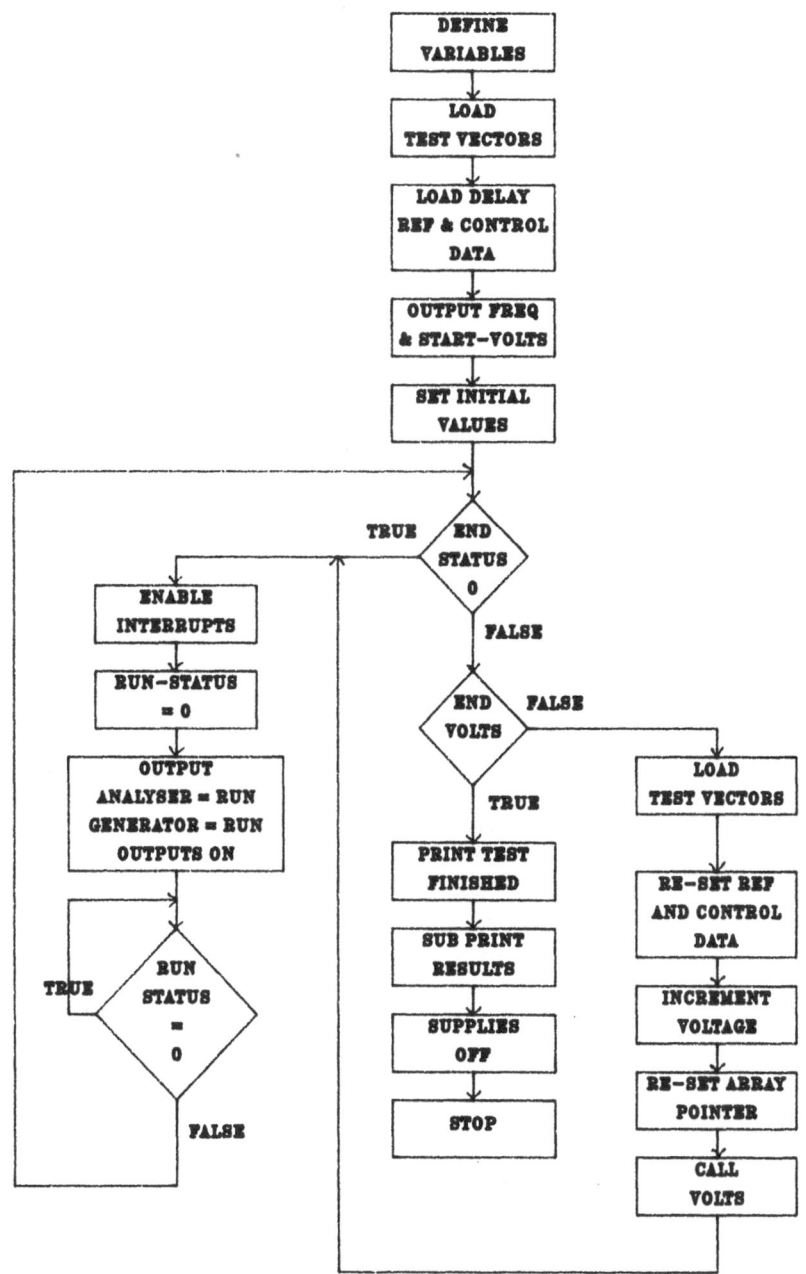

Fig. 5.4.   Flow diagram showing the procedures involved in the execution of the delay test on the multiplier chip.

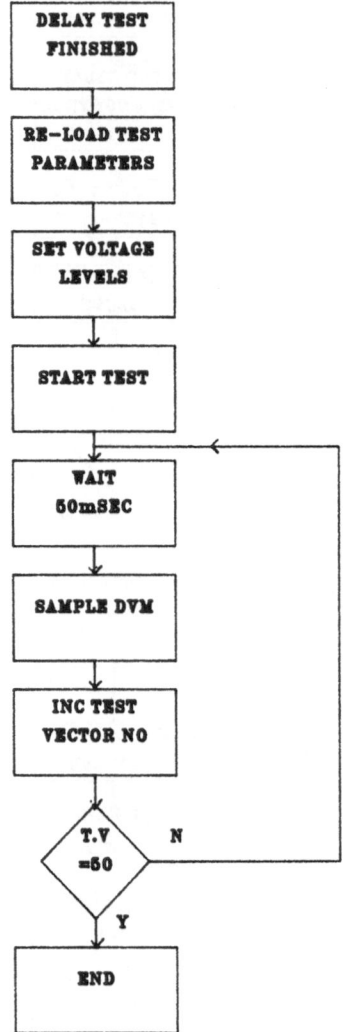

Fig. 5.5. Flow diagram showing the procedures involved in the execution of the StCT on the multiplier chip.

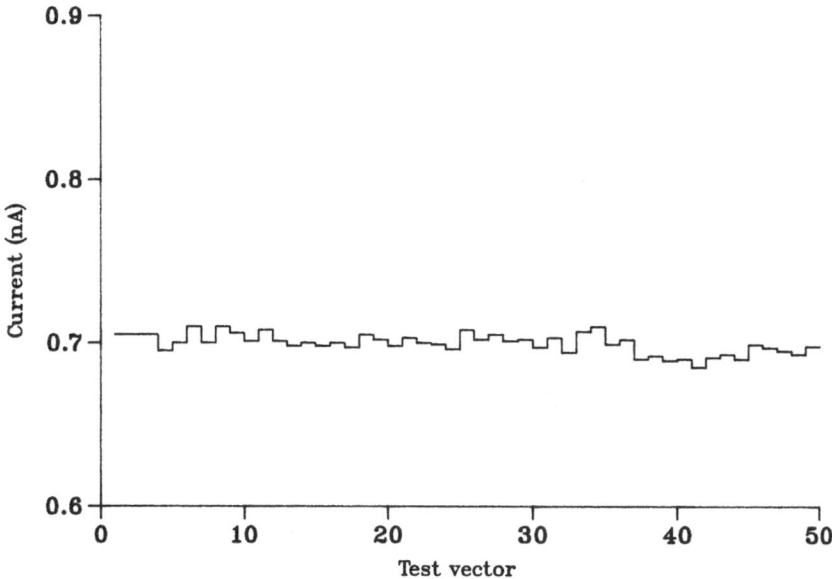

Fig. 5.6.   Results of the delay versus test vector on the multiplier reference device. Measurements were taken at 25^{0}C.

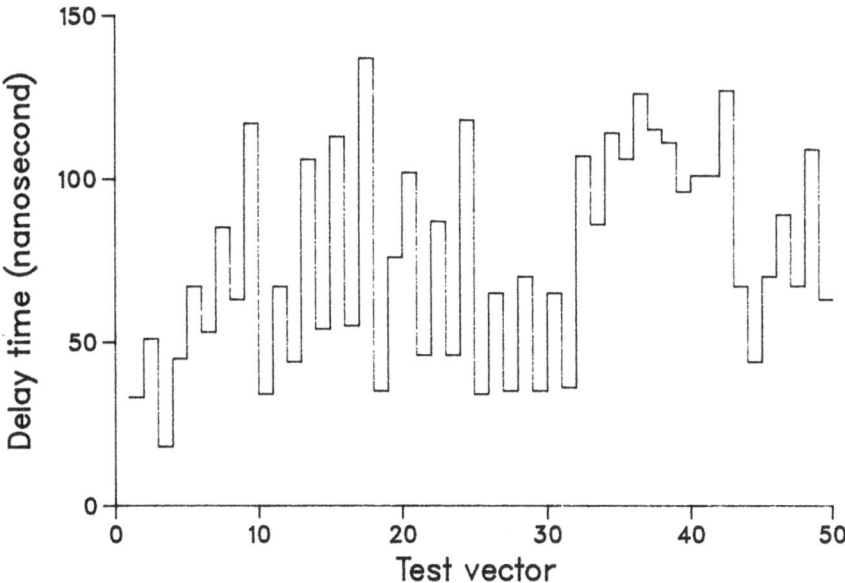

Fig. 5.7.   Results of the StCT versus test vector on the multiplier reference device. Measurements were taken at 25^{0}C.

different test vectors. This suggests that a leakage current threshold could well be applied to the leakage current test, whereas an comparative analysis would need to be performed for the delay test.

5.6. IMPLEMENTATION OF THE TESTS USING DEDICATED ANALOG CIRCUITS

5.6.1 Introduction

The experimental results described earlier were obtained using laboratory instruments which enable high quality measurements to be taken and are very flexible so that full research could be undertaken of the value of the tests. For most purposes in an industrial application it will be most cost-effective to use modified functional test equipment and this approach has been described in the previous section. However, in some cases there may be value either in using specialized, and hence expensive, test equipment or in developing more specialized dedicated test circuits. This approach will be outlined in this section.

Two basic approaches which can be used to simplify or speed up the tests will be given. These are sampling methods which use the power of current fast sampling equipment and signal processing, and the comparison methods which use apparatus designed specifically for one measurement and simplify the detection of any deviation from the normal by direct comparison of a measurement with the same measurement on a normal reference sample.

5.6.2. Sampling Methods

Sampling methods are of little use in speeding up the cut-off frequency test, CoFT. This test can only be performed on one test signal path through the circuit at a time. A more rapid measurement on each signal path can be accomplished by studying in detail the shape, rise-time and delay of the output signal after an input signal consisting of both a positive-going and a negative-going transient. This would be faster than the present method which observes several cycles of an input signal.

Although the other tests all measure the supply current under a variety of test vectors, they have to be considered separately in practice because of the several orders of magnitude difference in the currents involved. The static current measurements, StCT and ScNT, will typically involve nanoamperes to picoamperes while the transient tests, TrCT and TrNT, involve milliamperes. For each test a current-to-voltage converter is needed in the power supply lead. At moderate speeds a precise operational amplifier circuit is possible with either linear or logarithmic response as described in the previous section. For fast transient measurements of large currents a 50 Ω current sensing resistor may be used as the converter and the voltage which is fed directly into a high frequency voltage measuring circuit as described in Chapter 3 for the TrCT. There is therefore a basic design difference for the large, fast current circuits and the small current sensing circuits which usually have to be slower. However, both circuits can feed into the same voltage measurement circuit.

The threshold voltage test, ThVT, gives very specific information on a part of the circuit. It can be performed rapidly if a fast triangular input voltage is applied to the gate under test instead of the slow ramp used earlier. Since only the individual input circuits are tested, not much decrease in the total test time is achieved by speeding this test. However, this test can use the same equipment as other tests.

Although the static current measurements are taken of the current between the transient pulses, in practice a different set of test vectors would be needed. We can, however, consider these tests together as far as the signal acquisition and processing is concerned.

In each case we can construct and execute a suitable set and sequence of input test vectors and sample the voltages for each vector. For the ScNT a single sample is sufficient during the test period once the circuit has settled down, but for the TrNT a sufficient number of samples will be needed in order to define the pulse shape or perhaps the width of the pulse if only the propagation speed is needed.

In this case a fast transient sampling system and more probably a sampling oscilloscope technique on a repetitive signal will be needed so that each short-acquisition time sample is taken at a different part of the transient on successive repetition to build up the whole transient. For specific measurements, efficient sampling algorithms can be developed. In some measurements more repetition may be needed so that signal averaging techniques can be used to enhance the signal-to-noise ratio.

The advantage of such a sampling system is that it can be used directly to make a very rapid noise measurement for either the ScNT or TrNT. Consider the full test measurement pattern over the full test vector sequence for either test. If the measurement is repeated several times then the average of these repetitions gives a measure of the current with a good signal-to-noise ratio. If the mean square of the differences between each trace and the mean is computed then a measure of the noise is produced. The time taken for the measurement is the same as that of a noise measurement at a single test vector or single time on a transient pulse. By using a sampling system the noise measurement is made faster by a factor equal to the number of sampling points on the trace. The reason for this is that at each repetition all points on the trace are incorporated into the measurement rather than just one.

A physical justification for the noise measurement technique is needed. The noise is that of a fluctuating, time-varying resistance. The measured signal is that due to the fluctuation in the current flowing through this resistance at constant voltage bias. The fluctuations in resistance are unaffected by the presence or absence of the sensing current flowing. This procedure has been verified for the TrNT.

There are practical difficulties with these noise measurements. Noise signals have a Gaussian amplitude probability distribution about the mean value of the fluctuating current. This requires that the samples of the time-varying quantity must have high resolution over a wide amplitude range in order that the peaks are detected as well as many small amplitude fluctuations. A high-resolution sampling system is needed, usually better than 10 bits. In these measurements the problem is much worse than this since the average value of the static leakage current may vary by an order of magnitude even between devices nominally identical. For the transient pulses the structure of each pulse needs to be resolved in time to better than one-tenth of the pulse length and in amplitude by at least 8 bits for adequate definition and to accomodate difference between nominally identical devices and peaks of different sizes. Therefore an extremely high signal resolution is needed in the sampling system unless special auto-ranging precautions are taken.

5.6.3. Comparison Methods

The problems resulting from the wide dynamic range of analog measurements are great. In the static current tests, StCT and ScNT, the variation between a good and bad device of the same type was shown to be many orders of magnitude. This is not too serious for production purposes since a threshold can be inserted above which the device is rejected. However, considerable variation can be found between processing batches and a large difference between different types, technologies and manufactures. Little variation is expected for the ThVT but some moderate variation may be expected for the other tests. This variation could be very inconvenient in the design of the circuit if it is to be flexible especially if the large resolution needed for the noise test is also required.

One measurement design philosphy that has been widely adopted to eliminate problems with the wide variability between specimens is that of making a comparison between the device under test and a standard device. This can be performed either as a simple 'better than-' or 'worse than-' test or as a measurement of the magnitude and sign of the difference signal. This removes some of the problems associated with the large dynamic range of the measured variable since only the difference signal is detected, amplified and processed. Automatic noise measurements, for example, would be much easier to perform this way.

The ThVT is an absolute test that need not be considered for a differential measurement. The two static current tests, StCT and ScNT, benefit greatly from such a technique and would need a differential current to voltage converter of some type. Similar considerations apply for the transient current tests, TrCT and TrNT. The CoFT could be simplified with an input sweep frequency generator and either a digital or an analog difference detector on the output of the two circuits being compared, depending on whether a simple threshold, pass/fail test is adequate or whether more detail is needed. The difference signal could be used to investigate the relative rise times or relative delays in the output signal after an input transient or the shape of the difference output signal.

With an analog difference signal, quantitive difference signals may be obtained without the need for high-frequency, high-resolution sampling systems. For example, consider the analog difference signal between two transient current signals, one from the device under test and the other from a reference device. The average value of this signal over the whole test vector sequence, or the noise in the signal, would give an overall measure of the device quality. This could easily be used to detect a poor process variable since then each transient pulse of the bad device would be broadened. Since the transient charge or pulse area depends mainly on the capacitance, the relevant average would be the positive-only (say) difference signal. This gives no discrimination between different subsections of the circuit since each pulse is not separated. An intermediate step before complete sampling of the pulse shape would be to apply an integrating sample and hold operation over the total width of each transient pulse.

Reliability testing is likely to be needed for small numbers of a wide variety of devices. A major problem in the implementation of any test is that a new test vector set is needed for each device together with the corresponding truth table for the output. This may be a slow process. The total test time (test vector generation and actual test) may be reduced using a comparison method at the expense of the speed of the test. As with functional testing, a set of random digital signals may be applied to the inputs of both the device under test and a reference test. A digital or analog comperator is then placed at the output to detect significant departures from equality.

5.6.4. Modulation Methods For Diagnosis

The static current tests, StCT and ScNT, and the transient current tests, TrCT and TrNT, are valuable for determining which gate, transistor or circuit element is weak. By choosing suitable input vectors or sequences, one or a few nodes can be exercised and a defective element will produce a signal with high sensitivity.

The CoFT does not necessarily have this facility. Consider a serial-in serial-out shift-register. A functional test will only show that one element is faulty if the signal does not complete the path. An abnormal response in the CoFT will similarly just indicate that one element is weak, but again there is no direct information about the location of that element. The same will be true of paths set up in any circuit for the CoFT unless a suspect weak element can be included in several different test paths. If we return to the shift-register example we see that the problem is that the signal at the output of the weak element is healed and restored to a clean binary signal after passage through the following gate.

By making use of the marginal voltage test the location of the weak gate can be determined. Consider the shift-register operating at a low supply voltage,

V_{dd}, and a frequency which is just below the cut-off frequency so that the input signal propagates to the output. A reduction in V_{dd} by a small amount will take the weak element below its operating state under these conditions but leave the other elements operating normally.

A simple modulation of V_{dd} can therefore reveal the weak element. Consider a pulse on V_{dd} which reduces its value and lasts for one clock period. If a single 1 signal is propagated down the register and V_{dd} is pulsed downwards when the signal is passing through the weak element then the signal path will be broken. If the V_{dd} modulation is strobed along the register the weak element will be detected. This technique is totally digital in nature, except for the amplitude of the modulation, so that it is ideal for implementation on a digital tester.

The TrCT is powerful as a diagnostic tool. If a device has a weak component such that a gate fails to switch as the supply voltage is reduced or the frequency is increased, then the pulse on the TrCT signature will disappear. In the example of the shift-register, the time of the clock pulse at which the error occurs gives information about the location of the weak component. Other modulation techniques such as a linear ramp can also be used but the details will be rather specific to the application.

5.7. APPLICATION TO OTHER DIGITAL FAMILIES

The investigation and results presented here have been applicable to digital devices made by CMOS technology. Whilst this is a very common technology and likely to be widely used for sometime, we will discuss briefly the application of the test strategies and the particular tests to other process technologies. One of the reasons for the popularity of the CMOS process is that there is no large current flowing through the device when it is not switching. The leakage current that flows is the basis of our StCT and ScNT. Similarly, the power supply current pulses which pass through the switching sections of the circuit are the basis of the TrCT and TrNT. Other digital technologies do not show these features so that the tests cannot all be transferred directly.

Many features of the tests, the leakage currents and some components of the excess, $1/f$, noise are the result of the deterioration of p–n junctions. Much of this deterioration is known to occur at the intersection of the junction with the silicon surface and is due to the surface and oxide states at and near the silicon–silicon oxide interface. These phenomena occur in bipolar devices so that one would expect these basic failure indicates to exist in all technologies. The threshold voltage test is naturally specific to MOS processes, although variants of this test may be possible for bipolar circuits.

The basic philosophy behind the tests, that sensitive analog measurements are able to detect incipient failure in digital circuits, should hold for all circuits but the scope and type of test may differ. It is expected that in all cases the best strategy would still to perform the tests at an elevated temperature and during the burn-in period. There are possible departures from this general principle. For example, if a bipolar technology is degrading because of surface leakage currents (which are similar to MOS channels) then these can be accentuated by cooling the circuit since the bipolar transistor action is reduced while the MOS action has a very small temperature dependence.

Transistor–transistor logic (TTL) uses bipolar technology. The transistors and gates switch on and off but with many gates in a circuit there are likely to be many gates on at any particular state of the circuit. In the static state the gates that are on pass a current so that any leakage current through a gate in the off state will not be easily detectable. The static current tests, StCT and ScNT, are not therefore very sensitive. There is no transient current during the switching operation but there will be a difference in the current flowing for each internal state of the system and a change as a gate switches between its states. Some information should be obtainable from the current level as a function of the input test vector. However, the change in current in a defective gate will not be great so that the discrimination will not be high. Also the current will be dominated by the output transistors of each gate so that not all of the transistors are monitored equally. The highest sensitivity will be obtained by observing the current change between states or as a difference experiment between the device under test and a comparison device. The spectrum of the current pattern for a given test vector set has been used as a convenient quantity to monitor.

The speed of response test, the CoFT, is still applicable and could also be studied by observing the rise and delay times of the supply current steps during a switching operation. Possibly the differential of the supply current could play a similar role to that of the TrCP.

Emitter coupled logic (ECL) has features that make the tests less valuable. It is a constant current logic so that the total supply current stays at a constant, high value during any switching. This means that both the static and transient current tests have no value. Since the current passes through different transistors in the different logic states it may be possible to detect the $1/f$ noise generated in a defective device. The current source transistors do not change state during the switching operation. Again the CoFT may be used to detect weak devices.

The study has been to investigate digital circuits using analog tests. Analog ICs may be tested directly with analog tests to observe the departure of the specified parameters from the ideal.

REFERENCES

Ager, D.J., and Henderson, J.C., 1981, Proc. 19th Ann. Int. Rel. Phys. Symp., 139-48.

Ager, D.J., and Mylotte, P.S., 1977, Microelectron. and Reliab., 16: 679-87.

Crapuchettes, C., Proc. 1987 Int. Test Conf., 310-5.

Hamaguchi, S., and Ishikawa, K., 1983, *IEEE J. Solid-state Circuits*, **SC-18**: 409-13.

Hashizume, M., Yameda, K., Tamesada, T., and Kawakami, M., 1988, *Proc. 1988 Int. Test Conf.*, 374-80.

Keating, M., and Meyer, D., *Proc. 1987 Int. Test Conf.*, 316.

Malaiya, Y.K., *Proc. 1983 Int. Test Conf.*, 560-71.

Melia, A.J., 1978, *Electronics Lett.*, **14**: 434-6.

Lesser, D.J., and Shedletsky, J.J., 1980, *IEEE Trans. Computers*, **C-29**: 235-48.

Moltoft, J., 1980, *Microelectron. Reliab.***20**: 787-802.

Smith, G.L., *Proc. 1985 Int. Test Conf.*, 342-9.

Storey, T.M., and Barry, J.W., 1977, *Proc. 14th Design Automation Conference*, 492-4.

Wagner, K.D., *Proc. 1985 Int. Test Conf.*, 334-41.

Chapter 6

CONCLUSIONS

6.1. VALIDITY OF THE EXPERIMENTS

In this book we have described a set of analog measurements which may be used to test the quality of a digital IC. Although other tests are possible, and have been proposed, we have described and used a set of tests which are sensitive to degradation of the circuit quality, are convenient to use, give results which can be readily interpreted and have potential for incorporation into an industrial environment using standard equipment.

It has been demonstrated that these tests give an analog output which can be taken as a measure of the test. This output remains constant at some initial value while the device has normal performance, but starts to depart very much from this value as the device becomes of poor quality and eventually fails to function correctly.

As tests of quality, analog measurements are preferable to simple digital functional tests which just give a pass/fail output. The analog measurement can give an abnormal indication well before, often orders of magnitude before, the device fails a functional test. The functional test is used at present because it is convenient for the reason that it uses conventional digital test equipment. Also its value can be extended to become more discriminating by taking the device to some extreme performance condition, such as an elevated temperature or low supply voltage. We have shown that some of the analog tests can be developed to be run on digital testers and they can also benefit by change in the performance conditions.

The tests have been studied to ensure that they do measure reasonable defect properties and the defects have been simulated by a package, PSPICE, to confirm the experimental observations. The change in the device properties as the device degrades has been verified and the tests have been found to be sensitive to even slight degradation, much less than would cause the device properties to fall outside the device digital specifications.

It has been shown that these tests can be developed to be used on simpler equipment, on modified digital testers and they can be performed rapidly. Although the experimental studies have been performed on SSICs and MSICs, the basic methods of extending the techniques to VLSICs have been developed and a demonstration experiment performed.

The experimental study has been performed on CMOS devices and the first batch was encapsulated in plastic and subjected to high temperatures. This may not have produced typical degradation phenomena but the lifetimes of the devices were found to be extremely long when compared with other devices, in ceramic packages, which were found to fail more easily. The experimental procedure actually used may not be realistic. The aim was to detect a weak component within an IC which passed a functional test. Such devices are not readily available. The experimental methodology adopted was to take a batch of devices and subject them to the set of non-destructive tests. The batch was then subjected to a temperature–voltage stress for a fairly short time and then the batch was retested. This sequence was repeated until the tests showed indications of degradation, which usually became greater, and the device finally failed a digital functional test. It is possible that the defects produced by this procedure are not the same as these existing in a batch of devices immediately after manufacture. There is some evidence that the tests can detect processing defects since many samples showed a burn-in effect. The test measurement showed a variation from the initial value during the first few hours and days of stress. This variation could be of either sign. The value then stabilized for a long period of stress until the degradation appeared. It has also been observed that the test measurements can reveal various instabilities within the device. For example, the device properties varied over a period of about 10 hours after reaching room temperature after a stress. This stabilization or recovery effect was also observed in some devices which showed an abnormal behavior. They recovered their normal properties after a period of days at room temperature.

Many of the tests have been reported in a brief form in the literature. The value of the present study is that it is of large scale and shows a comparative evaluation of all the tests on many samples under different stresses. This comprehensive investigation has also included simulation of the devices. The results obtained have confirmed our understanding of the operation of the device and have illustrated qualitatively and quantitatively how the tests respond to different degradation modes. The simulations have shown that CMOS is a robust technology such that the operation of the circuit within specification is not critical on the exact process parameters.

The device performance is also not degraded rapidly by poor quality of the device either after processing or after degradation by stress. This indicates that our tests need not measure small changes in the device parameters, such as leakage current, since only large variations are of importance to produce a degraded performance. However, our experiments have shown that small changes are indicative of future large changes and later failure.

6.2. RECOMMENDED TESTS

Several tests appear to be of most value after this study. They have advantages and disadvantages and have different sensitivity to various defects. The choice will depend on the specific application.

The threshold voltage test, ThVT, measures the threshold voltage and gain factor of each of the transistors in the complementary CMOS pair at an input terminal. The test is rapid and a measurement at all inputs will give a statistical sample of the magnitude and variability of those quantities over the area of the chip. This gives information about the process accuracy but only with bias heat treatment or similar test can it help a reliability study.

The static current test, StCT, and the static current noise test, ScNT, investigate the magnitude and quality of the leakage current through the whole circuit when the cirucit is not operating. They mainly measure the deterioration or breakdown of the reverse biased p–n junctions in the circuit. Since junction leakage is a frequent result of defects, and the current can change by many orders of magnitude these are very valuable tests. Since different test vectors expose different components to a reverse voltage, the tests have great flexibility for a fast but coarse resolution test or a slower but more sensitive test. With a range of test vectors there is a possibility in principle of detecting the fault location, and from the full I~V characteristic an indication can be obtained of the fault mechanism. The tests can easily be implemented on a modified digital tester.

The noise in this leakage current is measured in the ScNT. Excess, or $1/f$ noise, is a very sensitive indicator of a defective component but since the noise is only large at very low frequencies, the measurement takes several seconds unless special multiplexing techniques are used.

The transient current test, TrCT, and the transient current noise test, TrNT, investigate the magnitude, shape and quality of the current which flows through the section of the circuit which is making a transition. Since only this subsection of the circuit is involved for each change between input test vectors, there can be high discrimination and sensitivity. Although the tests have great potential and, in theory, can reveal a large amount of information, they are not easy to perform and require very specialized equipment and experimental techniques since the transient pulses have a length comparable to the transition time of the digital circuit and the structure within these pulses contains the information. As circuits improve in speed of operation this test becomes more difficult to conduct. There will be problems in actually connecting the high frequency equipment to the device under test at these frequencies. These tests also have potential for fault location and diagnosis but the interpretation of the measurements is not easy at present. The noise test is also slow unless special techniques are used.

The cut-off frequency test, CoFT, also measures the speed of response of the circuit but is simpler and hence does not reveal so much information. It can often help to give information on the fault location plus it is very fast. By a study of the results of the test with power supply voltage, some diagnostic information is possible. This test is very easy to implement on a digital tester.

The recommended procedure is to perform the StCT and CoFT on a modified digital tester during the burn-in period. This is 'free' time and the effect of the deterioration of the device is enhanced and the current is larger and the frequency lower so that the measurements are easier.

6.3. USE OF THE TESTS

The tests give early warning of deterioration of various components in the circuit. These are mainly the junction leakage currents, threshold voltages, current-sourcing ability of the transistors and increase in the resistance of the interconnects. Many defects of a circuit can affect these or other quantities which can be detected. Routine tests of production devices will give a good indication of the actual value of the device parameters achieved. These are non-destructive and non-stressing tests so that these tests are suited better than special test circuits or drop-ins to monitor the process. This has implications for yield improvement. The tests are ideal for a given device process and manufacturer since statistical information can be easily built up on the test output parameters which can be achieved.

There is more difficulty in comparing the test results between processes, devices or manufacturers. However, statistical information can be built up rapidly and this can provide a standard against which defective devices can be rejected. For example, any normal device from the same source might be expected to be within one standard deviation of the test parameter from the average. Thus the tests may be used for incoming or outgoing quality assessment.

For some measurements, such as the leakage current, an absolute comparison between different suppliers may be expected to be indicative of the quality.

The aim of the investigation was to produce tests which could reduce, simplify and even replace some of the work done in producing high reliability devices. These non-destructive and non-stressing tests on the real devices might be able to replace much of the inspection and monitoring of special test chips in high reliability production. Each probe and test of the actual chip is likely to lead to possible deterioration. It is much better to test the finished device. Certainly visual inspection and other such tests will be necessary. It should be stressed that these tests cannot replace accelerated stress lifetime testing as a proof of a process technology. The parameters of the tests have no absolute values. If a batch of devices has been produced the tests can detect deviant devices, and if there is a choice then those devices at the higher quality end of the distribution of the test parameter over the batch may be chosen for high reliability use.

6.4. FURTHER WORK

Although this work has attempted to be complete, reliability testing is a very lengthy study and a time has to be chosen for publication of the results. Much work could be done to extend the study.

The initial batch studied was of 4013 devices in plastic encapsulation. This was unfortunate and may have led to untypical results. A detailed study should be made of devices in ceramic packages. The devices used have been commercial devices because these would be expected to produce a wide range of quality and a large defect rate. For a valid test of the ideas behind this work a comparison should be made between the same devices selected into different grades of reliability. The suggestion that the tests are best made during the burn-in procedure was made after analysis of the results from Batch 1. This suggestion needs practical verification.

A basic problem with the investigation is the assumption that the experimental procedure of stressing the devices to produce defective devices actually produces devices with defects typical of processed, but unstressed, devices. This assumption needs investigation. A large batch of new devices could be taken and tested before installation in a life test or in an actual system. The results of the tests would then indicate which devices would be likely to fail early in their life. The results would be interesting but would be likely to take some time to appear.

The transient current test shows considerable substructure and must contain much information about the detailed operation of the device. A further investigation of its characteristics would probably be fruitful. Although the equipment needs to be very specialized it is likely that sampling systems faster than the fastest digital circuit will always be available. The test is only suitable for a laboratory environment.

It is apparent that similar techniques could be applied to GaAs digital circuits once standard configurations have been established. The low frequency noise is large and characteristic and the response times due to the trapping states are characteristic and measurable from the slow component of switching transients.

Appendix 1

SPICE CIRCUIT FILES USED
IN THE SIMULATIONS

1) The 4013 D-type flip-flop

FLIP-FLOP (CMOS)

    * DEFINE TRANSISTORS

    .MODEL MN NMOS (LEVEL=1 VTO=1.7V KP=23U VMAX=1E5 IS=1E-11)
    .MODEL MP PMOS (LEVEL=1 VTO=-1.2V KP=15U VMAX=1E5 IS=1E-11)
    .MODEL MN1 NMOS (LEVEL=1 VTO=1.7V KP=2.2U VMAX=1E5 IS=1E-11)
    .MODEL MP1 PMOS (LEVEL=1 VTO=-1.2V KP=1.5U VMAX=1E5 IS=1E-11)

    * SPECIFY INPUTS

    VDD   3 0 DC 5.0
    VDATA   10 0 PULSE (0V 5.0V 995N 5N 5N 995N 2000N)
    VCLIN   20 0 PULSE (0V 5.0V 495N 5N 5N 495N 1000N)
    VSET   12 0 DC 0V
    VRESET 14 0 DC 0V

    * NOR GATE

    .SUBCKT NOR 1 3 4 2
    MP1 5 3 2 2 MP L=9U W=80U
    MP2 4 1 5 2 MP L=9U W=80U
    MN1 4 3 0 0 MN L=9U W=40U
    MN2 4 1 0 0 MN L=9U W=40U
    C1   5 0 0.06P
    C2   3 0 2.6P
    C3   1 0 2.6P
    .ENDS NOR

```
.SUBCKT TGATE 1 4 5 6 2
MP3   1 5 4 2 MP L=9U W=80U
MN3   1 6 4 0 MN L=9U W=80U
C10   1 5 0.12P
C11   1 6 0.12P
C12   5 4 0.12P
C13   6 4 0.12P
.ENDS TGATE

.SUBCKT INVA 1 4 2
MP4   4 1 2 2 MP L=9U W=40U
MN4   4 1 0 0 MN L=9U W=40U
C4    1 0 1.4P
.ENDS INVA

.SUBCKT INVC1 1 4 2
MP4   4 1 2 2 MP1 L=9U W=150U
MN4   4 1 0 0 MN1 L=9U W=100U
C8    1 0 6.2P
.ENDS INVC1

.SUBCKT INVB 1 4 2
MP5   4 1 2 2 MP L=9U W=100U
MN5   4 1 0 0 MN L=9U W=100U
C5    1 0 3.2P
.ENDS INVB

.SUBCKT INVC 1 4 2
MP6   4 1 2 2 MP L=9U W=150U
MN6   4 1 0 0 MN L=9U W=100U
C6    1 0 6.2P
.ENDS INVC

.SUBCKT INVD 1 4 2
MP7   4 1 2 2 MP L=9U W=400U
MN7   4 1 0 0 MN L=9U W=160U
C7    1 0 14P
.ENDS INVD

* NETWORK
* CLOCKS

X1  20 21 2 INVA
X2  21 22 2 INVB
```

```
X3   22 23 2 INVB

* MASTER

X4   10 11 22 23 2 TGATE
X5   11 12 13 2    NOR
X6   13 14 15 2    NOR
X7   15 11 23 22 2   TGATE

* SLAVE

X8   13 16 23 22 2 TGATE
X9   16 14 18 2    NOR
X10  18 12 19 2    NOR
X11  19 16 22 23 2 TGATE

* OUTPUT BUFFERS

X12 18 31 2 INVA
X13 31 32 2 INVC
X14 32 33 2 INVD
X15 16 34 2 INVA
X16 34 35 2 INVC
X17 35 36 2 INVD

* OUTPUT CAPACITORS

C33   33 0 .1P
C36   36 0 .1P

* POWER SUPPLY RESISTOR

RLOAD 3 2 4

* SET INITIAL VALUES

.NODESET V(21)=5.0 V(22)=0.0 V(23)=5.0 V(11)=0 V(13)=5.0
+V(15)=0 V(16)=5.0 V(18)=0 V(19)=5.0 V(34)=0 V(35)=5.0
+V(36)=0 V(31)=5.0 V(32)=0 V(33)=5.0

.TRAN 50N 3000N
.OPTION NODE TNOM=22 ITL4=20 LIMPTS=401 ACCT
.PROBE
.END
```

2) The 4014 8-bit shift-register

4014 cutoff frequency test on 4014 stage 1 plus buffers Run 02

```
    * DEFINE TRANSISTORS

    .MODEL MN NMOS (LEVEL=1 VTO=0.8V KP=18U TOX=1000N UO=508
VMAX=1E5)
    .MODEL MP PMOS (LEVEL=1 VTO=-0.8V KP=9U TOX=1000N UO=254
VMAX=1E5)

    * SPECIFY INPUTS

    VDD    1 0 DC 5
    VPS    2 0 DC 0
    VSIN   4 0 PULSE (0V 5V 50N 10N 10N 60N 140N)
    VCL    3 0 PULSE (0V 5V 5N 10N 10N 25N 70N)
    VPIN1 14 0 DC 0

    * NAND GATE

    .SUBCKT NAND 1 3 2
    MP1  3 2 2 2 MP L=9U W=160U
    MP2  3 1 2 2 MP L=9U W=160U
    MN1  3 1 4 0 MN L=9U W=280U
    MN2  4 2 0 0 MN L=9U W=280U
    C1   4 0 0.05P
    C2   1 0 2.24P
    .ENDS NAND

    .SUBCKT TGATEA 1 2 3 4 5
    MP3  1 3 2 5 MP L=9U W=20U
    MN3  1 4 2 0 MN L=9U W=16U
    C3   3 0 0.1P
    C4   4 0 0.1P
    .ENDS TGATEA

    .SUBCKT TGATEB 1 2 3 4 5
    MP4  1 3 2 5 MP L=9U W=40U
    MN4  1 4 2 0 MN L=9U W=30U
    C5   3 0 0.2P
    C6   4 0 0.16P
    .ENDS TGATEB
```

```
.SUBCKT TGATEC 1 2 3 4 5
MP5   1 3 2 5  MP L=9U W=12U
MN5   1 4 2 0  MN L=9U W=12U
C7    3 0 0.06P
C8    4 0 0.06P
.ENDS TGATEC

.SUBCKT NOTA 1 2 3
MP6   2 1 3 3  MP L=9U W=20U
MN6   2 1 0 0  MN L=9U W=20U
C2    1 0 0.2P
.ENDS NOTA

.SUBCKT NOTB 1 2 3
MP7   2 1 3 3  MP L=9U W=20U
MN7   2 1 0 0  MN L=9U W=16U
C3    1 0 0.2P
.ENDS NOTB

.SUBCKT NOTB1 1 2 3
MP21  2 1 3 3  MP L=9U W=20U
MN21  2 1 4 0  MN L=9U W=16U
R1    4 0 1000
C11   1 0 0.2P
.ENDS NOTB1

.SUBCKT NOTC 1 2 3
MP8   2 1 3 3  MP L=9U W=40U
MN8   2 1 0 0  MN L=9U W=30U
C4    1 0 0.34P
.ENDS NOTC

.SUBCKT NOTC1 1 2 3
MP20  2 1 3 3  MPF L=9U W=40U
MN20  2 1 0 0  MN L=9U W=30U
C10   1 0 0.34P
.ENDS NOTC1

.SUBCKT NOTD 1 2 3
MP9   2 1 3 3  MP L=9U W=12U
MN9   2 1 0 0  MN L=9U W=12U
C5    1 0 0.12P
.ENDS NOTD
```

```
.SUBCKT NOTE 1 2 3
MP10 2 1 3 3 MP L=9U W=50U
MN10 2 1 0 0 MN L=9U W=20U
C6    1 0 .36P
.ENDS NOTE

.SUBCKT NOTF 1 2 3
MP11 2 1 3 3 MP L=9U W=400U
MN11 2 1 0 0 MN L=9U W=160U
C7   1 0 2.86P
.ENDS NOTF

.SUBCKT NOTG 1 2 3
MP12 2 1 3 3 MP L=9U W=64U
MN12 2 1 0 0 MN L=9U W=64U
C8    1 0 .64P
.ENDS NOTG

.SUBCKT NOTH 1 2 3
MP13 2 1 3 3 MP L=9U W=160U
MN13 2 1 0 0 MN L=9U W=160U
C9    1 0 1.62P
.ENDS NOTH

* NETWORK

* PS

X1   2 5 1 NOTG
X2   5 6 1 NOTH

* CLOCK

X3   3 7 1 NOTG
X4   7 8 1 NAND

* DIN

X5   4 9 1 NOTA
X6   9 11 1 NOTB

* STAGE 1

X7   11 15 6 12 1 TGATEA
```

```
X8    14  15  12  6  1  TGATEA
X9    15  16  8  13  1  TGATEB
X10   16  17  1  NOTC
X11   17  18  1  NOTB
X12   18  16  13  8  1  TGATEB
X13   17  19  13  8  1  TGATEB
X14   19  21  1  NOTB
X15   21  20  1  NOTD
X16   20  19  8  13  1  TGATEC
X17   6  12  1  NOTB
X18   8  13  1  NOTB

* STAGE 2 ~ 8

X19   6  22  1  NOTB
X20   8  23  1  NOTB
X21   6  32  1  NOTB
X22   8  33  1  NOTB
X23   6  42  1  NOTB
X24   8  43  1  NOTB
X25   6  52  1  NOTB
X26   8  53  1  NOTB
X27   6  54  1  NOTB
X28   8  55  1  NOTB
X29   6  56  1  NOTB
X30   8  57  1  NOTB
X31   6  58  1  NOTB
X32   8  59  1  NOTB

* BUFFERS

X33   21  60  1  NOTE
X34   60  61  1  NOTF
C10   61  0  3.75P

* SET INITIAL VALUES
.NODESET V(16)=0 V(19)=5 V(5)=5 V(6)=0 V(7)=5 V(8)=0 V(12)=5
+V(13)=5 V(9)=5 V(11)=0 V(15)=0 V(17)=5 V(18)=0 V(21)=0 V(20)=5

* COMMANDS
.TRAN 10N 450N
.OPTION NODE ACCT TNOM=22 RELTOL=.01 ITL4=40 LIMPTS=1000
ITL5=0
```

.PROBE V(3) V(4) V(8) V(11) V(13) V(15) V(16) V(19) V(21) V(17) V(60) V(61)
    .END

3) Half-adder from the semi-custom multiplier chip

MULTIPLIER (HALF ADDER 4)

* READ TRANSISTOR DEFINITION FROM LIBRARY

.LIB MCE.LIB

* SPECIFY INPUTS

```
VDD   3 0  DC  1.3
VX0   10 0  AC  1.3 PULSE(0V 1.3V 1U 20N 20N 1.96U 4U)
VY0   11 0  DC  1.3 PULSE(0V 1.3V 1U 20N 20N 1.96U 4U)
VCIN  12 0  DC  0
```

* DEFINE SUBCIRCUITS

* NAND GATE

```
.SUBCKT NAND 3 4 5 6
MP1  6 4 3 3 MP
MP2  6 5 3 3 MP
MN1  6 4 7 7 MN
MN2  7 5 0 0 MN
.ENDS NAND
```

* NOR GATE

```
.SUBCKT NOR 3 4 5 6
MP3  7 4 3 3 MP
MP4  6 5 7 7 MP
MN3  6 4 0 0 MN
MN4  6 5 0 0 MN
.ENDS NOR
```

* INVERTER

```
.SUBCKT INV 3 4 5
MP5  5 4 3 3 MP
MN5  5 4 0 0 MN
```

197

```
.ENDS INV

* OUTPUT BUFFER

.SUBCKT INVO  3  4  5
MP5   5  4  3  3  MPO
MN5   5  4  0  0  MNO
.ENDS INVO

* TEMP INVERTER

.SUBCKT INV1  3  4  5
MP5   5  4  3  3  MP1
MN5   5  4  0  0  MN1
* RLK  4  0  10MEG
.ENDS INV1

* NETWORK

X1   3  10  11  20  NAND
X2   3  20  21      INV
X3   3  21  12  22  NAND
X4   3  21  12  23  NOR
X5   3  23  24      INV
X6   3  22  30      INV
X7   3  22  24  25  NAND
X8   3  25  31      INV
X9   3  31  32      INVO

* OUTPUT CAPACITORS

CCOUT  30  0  1P
CTRA1  31  0  .26P
CIN1   10  0  1P
CIN2   11  0  1P
CIN3   12  0  1P
CBUF   32  0  15P

* ANALYSIS

.OP

* .AC DEC 5 100K 10MEG
```

```
.TRAN 40N 2U

* OUTPUT

.OPTIONS NODE NOPAGE NOECHO ACCT NOMOD ITL4=40
+RELTOL=.01

* .PLOT TRAN V(10) V(20) V(21) V(22)
.PROBE

.END
```

INDEX